Animal Choice and Human Freedom

Animal Choice and Human Freedom

On the Genealogy of Self-Determined Action

Michael Yudanin

LEXINGTON BOOKS
Lanham • Boulder • New York • London

Published by Lexington Books
An imprint of The Rowman & Littlefield Publishing Group, Inc.
4501 Forbes Boulevard, Suite 200, Lanham, Maryland 20706
www.rowman.com

6 Tinworth Street, London SE11 5AL, United Kingdom

British Library Cataloguing in Publication Information Available

Library of Congress Cataloging-in-Publication Data Available

ISBN: 978-1-7936-2018-7 (cloth)
ISBN: 978-1-7936-2019-4 (electronic)

Contents

Acknowledgements

This book is but a tiny island in a veritable archipelago of human thought attempting to understand freedom. As such, it had been formed by innumerable currents, some of which I would like to acknowledge specifically.

I owe the opening of my mind to a literature teacher in the far-away 1980s Novosibirsk, Elvira Nikolaevna Goryuchkina. While many teachers see their students' minds as a vessel to fill up with a variety of contents, she considered ours as a torch to lit and excite with the quest for the betterment of yourself and humanity. Without her, I would have never thought of stepping on the rough road of freedom, political as academic.

My quest for liberty not always had a scholarly character. These were my parents who endured its vicissitudes, and to them I owe a great debt which I probably will never repay.

This book would have never come into existence without philosophers who had crucial influence on my thinking. The scholarship of Richard Dien Winfield was invaluable. His books were enlightening to me and to many other philosophers. I was very fortunate to enjoy his guidance. Edward C. Halper had a formative effect on my thinking, particularly in the areas of philosophical methodology, philosophy of mind and emotion, and Aristotle. Piers H. G. Stephens encouraged me to pursue the study of freedom, provided excellent suggestions regarding the background reading, and opened before me the fascinating world of the philosophy of nature.

I wish to thank my dear friend Kirill Voytsel. His brilliant mind always encouraged me to try harder, and his insightful comments on the first drafts of this book were extremely valuable.

Irena, the love of my life and my best friend, thank you for enduring me for so long. It is standing by your side that gives me the power to move forward and do it better under your encouragingly critical gaze.

My dear daughters, Vered and Elizabeth, we are leaving a pretty messed up world to you. It will be up to you to hold high the banner of freedom.

Most importantly, I want to thank all the animals in my life, and particularly Sammy, Lily, and Tiger. I treasure your choice to provide me with your unconditional love.

Introduction

The Question of Freedom

FREEDOM AS A SPECIAL CONCEPT: THE UNITY OF THE HOW AND THE WHAT OF FREEDOM

Freedom is a natural subject for philosophy. The word expresses a concept that is abstract, devoid of any sensory content: it does not represent either a specific or a generalized image. Yet freedom is also something very worldly. It is more than heavily used in a variety of contexts and is related to a spectrum of unmistakable and related meanings in a plethora of situations, personal as well as political. The concept of freedom, through a variety of its conceptions, plays a pivotal role in the design of various systems of government, motivates massive political movements, and is being used as a justification by innumerable individuals trying to explain their decisions and actions.

Few would fail to understand the use of 'freedom' in the contexts of slaves proclaiming their desire to be released form bondage, the laments of a teenager about the restrictions imposed by her parents, when asserting the ability to follow whatever alternative one prefers, as related to releasing animals from confinement, or in the 'degree of freedom' when addressing the range within which a certain parameter might fluctuate. All the uses of this term, except the metaphorical one as in the last example, pertain to the condition of agents in their relation to the world as being able to behave in it as they please; the limits of freedom address the degree to which the agents can do so. Since the core of the concept refers, however loosely, to the ability of agents to determine their own behavior, its very essence is self-determination. Freedom is about defining itself, it is its own starting point. The concept of free action is about being one's own cause, not being caused by an external factor. This makes it similar to the original condition of Hegel's Spirit, the beginning of reasoning – and, therefore, the conceptual birthplace of philosophy, a discipline that creates its own foundations. At the same time, this suggests that a philosophical inquiry into the concept of freedom cannot assume that freedom simply refers to a certain condition as, for example, the concepts of mind or politics are assumed to refer – freedom is about defining itself, a free entity is an entity that determines itself.

As such, the concept of freedom cannot have any externally given content – otherwise it will not be free, will not be self-determining. This has a number of consequences which will be discussed throughout this book. However, the one that is crucial for understanding why most of the debates on freedom have led to the dead-end of a battle of definitions is that for a concept that is about defining itself, the descriptive and the functional aspects collapse: to explain *what* freedom is is to describe *how* it is possible, how it can be carried out by agents that are purportedly free. For other concepts that claim to have an external referent, e.g., 'tree,' 'idea,' 'mind,' or 'epistemology,' these two aspects are separate. One can explain what the tree is by defining it and thus providing an idea of what sort of phenomena would fall under the concept of a tree. Explaining how the concept is instantiated is a different enterprise that, once we understand the concept of a tree, can be carried out by botanists, for example. The concept itself might be clarified based on their feedback, yet the separation between the descriptive and the functional aspects still stands. The same applies to 'epistemology': defining it as a philosophical discipline that pertains to the ways of knowing and figuring out whether knowing is possible or how it can be carried out are two distinct endeavors. Even if epistemological exploration will discover that knowledge is impossible and the word 'knowledge' is empty of meaning, the term 'epistemology' will still make sense as referring to this specific failed undertaking.[1] This approach does not work with freedom. Since it is not about something else but about itself, we cannot define it first and then inquire into how it works: the way it works is what it is, and what it is amounts to nothing more and nothing less than the way it works. If we separate these two aspects, then we will have no way to adjudicate between the different conceptions of freedom and will get mired in a war of definitions – which is precisely the state of the freedom debate for the last few decades.[2]

Frustrated with the unyielding nature of freedom, several philosophers have suggested that comprehending its meaning is beyond human understanding. Kant, for example, considered freedom to be "an inscrutable faculty which no experience could prove."[3] More recently, Noam Chomsky hinted that free will might be beyond the limits of our comprehension.[4] This seems, however, misguided. We feel that we do have freedom to choose in many, if not most situations, and that others like us have it too. This feeling, this perception that is cognitive rather than sensory, not unlike Descartes's "clear and distinct perceptions,"[5] seems to be fundamental to our social functioning. Our moral judgment, planning of actions involving more than one person, judiciary – pretty much every aspect of life that involves human beings and, in some cases, other mammals – assume freedom to choose what to do, implicitly or explicitly. This alone would justify attempting a serious exploration of freedom. Moreover, any layman can provide a commonsensical account of freedom: being able to do what you wish. A bit more thinking would yield a conclusion that the ability here is limited by various circumstances, yet this would not eliminate the fact that a basic account of freedom can be given quite easily. This is different from the cases of many other concepts, e.g., those of infinity, cardinality, space-time, and multi-dimensional universe. In all those cases a coherent verbal account is not easy for a non-expert to provide, and yet none of them is usually considered intractable.

The problems start cropping up when we try to flesh out the meaning of being able to do what you want. Can we ever do what we want, given that we live inside a world of physical, social, and other constraints? Can we control what wishes we have, given that we are frequently driven by the urges that are rooted in our physiology, e.g., hunger or empathy? Even the repertoire of our desires that do not seem to be driven by our genetic endowment is limited – a person who does not know anything about Jackson, Missouri, cannot wish to travel there. Finally, we might wish for two opposite things at once, for example, to steal a cookie to satisfy a genuine desire to eat it and not to steal a cookie to satisfy a no less genuine desire not to steal that follows from a deeply held conviction that stealing is wrong, a conviction that elicits strong emotional reactions. How is a coherent conception of freedom possible, given all these issues?

Philosophers usually concentrate on answering these questions by providing a clever definition of freedom that would satisfy as many of them as possible, and then focus on trying to render the other questions meaningless. Alternatively, some embrace the issues and argue that freedom is indeed an illusion. The subsequent debate frequently turns into a battle of definitions, where different sides are trying to poke holes in the definition of freedom provided by their opponents. A typical case here is the debate between compatibilists and incompatibilists.[6] Within the incompatibilist camp, a no less active discussion is raging between the libertarians who argue that some sort of absolute freedom of choice is possible in embodied beings and the determinists who deny that physical beings can have even a grain of absolutely free choice. Both incompatibilist camps are at odds with the compatibilists who are trying to provide a definition of freedom that accommodates the deterministic nature of physical reality. All sides supply numerous definitions of freedom, some less susceptible to the opposition's critique than others. Yet no camp that argues for the possibility of freedom is making serious attempts to explain *how* their version of freedom can be instantiated in the allegedly free agents and how it came to be in terms of the development of these agents' biology and cognition.

What eludes us is not necessarily the *what* of freedom but the *how* of it, the flesh on the skeleton provided by the most plausible model. This is crucial, as the *what* and the *how* of freedom are one and the same thing. This book will attempt to shed a light on the *how* of freedom, the details of how freedom as self-determination in the world is possible. Specifically, I will highlight one of the more important aspects of understanding freedom: the nature of self-determined choice and its relation to cognition, and thus to the life forms of the agents who are allegedly free. To do that, I will try and explore the most plausible conception of freedom given the way it can be instantiated in the world of material agents, and what capacities are required so they can exercise freedom. The latter, if we are considering biological agents, necessitates exploring freedom's phylogenetic development, explaining how the capacities that underlie freedom evolved in the course of the development of species.

FREEDOM AS A MULTIFACETED PHENOMENON

Bookchin claims that the term 'freedom' is absent in preliterate societies.[7] While this statement is hard to substantiate, any major language in use today has a term for freedom, and these terms can be relatively easily translated to other languages. This points to the universality of the condition to which the concept relates. Even more interestingly, the etymology of the terms used to express the concept of freedom in various languages points to different aspects of the use of the concept. In all Slavic languages the word for freedom is a variation on *svoboda* (свобода).[8] It has the Indo-European root *sva*, which means 'own' and is used in words like *svoj* (one's own); the Sanskrit *swatantra* (स्वतन्त्र), 'freedom,'[9] literally 'own doctrine,' and *swaraj* (स्वराज्), 'self-rule,' use the same root. Similarly, the Chinese word for freedom, *ziyou* (自由), consists of 'self' and 'by the means of,' or 'cause.' Here the aspect of self-determination, as an important element of the way the term is used, is emphasized. However, many of the Slavic languages also have a variation of *wolja* (воля) which also means freedom yet is usually used in less formal contexts, e.g., when talking about animals being unrestrained in their natural environment as opposed to being confined in cages. This clearly relates to the Germanic 'will' and, in fact, in some languages, like Russian, is also used to refer to the agent's will. The English 'freedom,' with its Germanic sisters like the German *Freiheit*, is usually traced to the Proto-Indo-European *pri*, 'to love,' which is still in use in this sense in Sanskrit, Hindi, and Marathi; the hypothesis here is that it came to designate freedom through 'beloved' or 'friend,' one of our own, a free person. Persian *azadi* (ازادی), which is also used in a variety of languages, e.g., Kurdish, Armenian, and Kashmiri, is rooted in the Proto-Indo-European *genh*, 'to be born into a clan,' and, by extension, 'noble' is of a similar etymology. In the modern Semitic languages, the ancient root *hrr* gives rise to the Hebrew *herut* (חירות) and Arabic *huriya* (حرية); its etymology is uncertain, and in all extant texts it always meant free as opposed to enslaved. Yet Hebrew also has *hofesh* (חופש), of uncertain etymology, and *dror* (דרור), which is related to ancient Akkadian, both meaning 'freedom' – which suggests that they descended from different words that probably addressed different aspects of freedom. Finally, the English 'liberty,' with its Latin roots that gave rise to the word in Romance languages, originates in *liber*, 'free,' itself having the origin in the Indo-European *lewd*, meaning 'people,' from which the Greek ἐλεύθερος originates as well. This hints at the understanding of freedom as something that is characteristic of humans. As such, it relates the oldest use of the word 'freedom' in writing, around 2350 BC in the Babylonian city of Lagash. In the Lagash tablets the word *amargi* is used to refer to the deliverance of the people from the oppressive rules of abusive elites; later, the same word has been widely used to designate the condition of a free as opposed to an enslaved person. The literal meaning of *amargi* is 'return to mother'[10] – perhaps, as Bookchin suggests, to the natural condition of humankind.[11]

This little etymological survey suggests the multiplicity of the aspects of the concept of freedom as it has been used since the dawn of history. Any serious exploration of freedom has to address it as a phenomenon. In other words, philosophical investigation of the concept of freedom should address the different aspects of freedom that are highlighted by the different uses of the term, try and test different conceptions of freedom against such uses and explore their legitimacy.

THE INVESTIGATION

To try and provide an account of the how of freedom that is possible for material agents, the investigation will proceed in three steps. First, I will explore the basic characterization of freedom. Then, I will develop a descriptive account of how freedom can be exercised in our material world, examining the different types of its instantiation, specifically distinguishing those in non-verbal animals capable of choice and in discursively intelligent humans. Lastly, I will investigate the evolution of choice as underlied by the biological evolution and proceeding in interaction with it.

Part I of the book is dedicated to exploring the concept of freedom and its possible forms. Chapter 1 will investigate the basic characterization of freedom. It will start with the common meaning of the concept and then extract three aspects of this account: the abilities that underlie it, the action that is necessary for it to be meaningful, and the self that wields freedom. Then, I will explore the necessary characteristics of a free entity, or the desiderata of freedom. I will argue for five such desiderata: access to realizable alternatives, imagination, activity in the world, self-control, and conscious intentionality. I will briefly address other criteria that are frequently suggested in the literature, and try to establish why the criteria of being able to act otherwise, unpredictability of action, and rationally made decision should not be necessary for defining a free entity. Based on the established desiderata, I will outline the specifications of a free self and its world. At this point, we will proceed to address the central problem of freedom – the seeming contradiction between free choice and determinism. We will examine this problem not as hanging in a philosophical vacuum but through the rich characterization of freedom provided in the preceding sections. Then, I will briefly discuss the main accounts of freedom. First, libertarian freedom, which argues that the only meaningful freedom is absolute, unlimited power of choice, and that a grain of such choice is present in free agents. Second, the deterministic approach that agrees that the only meaningful account of freedom is that of absolutely free choice, yet argues that such freedom is impossible. Lastly, I will explore the compatibilist account. It agrees with the determinists that absolute freedom is impossible in material beings, yet argues that on a more humble definition of freedom, that of freedom limited to choosing between available alternatives, some material agents can be free. I will suggest that the compatibilist account is the only account of freedom that is feasible, given the physical nature of the entities that can be thought of as free in our world. Then I will outline the problem of this account: the lack of adequate specifications of how the compatibilist freedom is possible in our world. The rest of this book is dedicated to an attempt to solve this problem by exploring the descriptive and the evolutionary aspects of compatibilist freedom.

Chapter 2 provides a differential characterization of freedom. It starts by suggesting that life is a phenomenon that answers the core requirement of freedom – self-determination. Living entities constitute meaningful selves that are functionally separate from their environment and carry the factors determining their activities and development within their bodies. Thus, if there is freedom, it will be instantiated in biological entities. Then, I will proceed to explore the different forms of life and discuss whether they demonstrate the abilities that answer the desiderata of freedom. I will argue that plants cannot be free, while animals can – and discuss the different abilities animals possess and their relation to choice. Exploring awareness, sentient consciousness, and representing consciousness, I will argue that the latter is the birthplace of freedom. After that I will address instincts and their relation to choice, and argue that instincts do not preclude choice but necessitate some degree of it. Before proceeding further, I will characterize animal choice and its limitations – this will be a segue to the discussion of human freedom. Addressing the latter, I will investigate how language and discursive intelligence are its defining characteristics. This will be followed by a characterization of human freedom, contrasting it with animal choice.

Part II will address the last crucial aspect of the *how* of freedom – its evolution. If, as Chapter 2 argues, some animals, including humans, are capable of self-determined choice, we should be able to trace the phylogenetic development of this ability. To do so, I will start with a brief discussion of life and evolution in Chapter 3, arguing, following Hans Jonas, that living entities cannot but evolve. Then, in Chapter 4, I will address those principles of biological evolution that are relevant to the discussion of freedom, and specifically variation, heredity, natural selection, and increase of complexity. I will try to establish that biological evolution is characterized by directionality toward an increase in complexity, and that this is likely to lead eventually to the development of animal choice and human freedom. In this way, biological evolution is somewhat similar to Hegel's unfolding of Spirit. While Hegel rejected the theories of biological evolution with which he had been familiar, I will argue that his account is evolutionary in nature. With that, in Chapter 5 I will proceed to build an account of biological evolution as the unfolding of freedom, providing the evolutionary aspect of the descriptive differential account outlined in Chapter 2. I will start with addressing the development of unicellular life, continue with the evolution of multicellular organisms, and argue that plants constitute a branch of the tree of life that exemplifies a dead end in terms of the evolution of freedom. Then I will address in more detail the evolution of non-verbal animals, emphasizing how the representing consciousness had evolved before the evolution of freedom could proceed any further. I will also address the limitations of animal choice that are imposed by its non-verbal nature. Finally, I will provide an account of the evolution of discursive human freedom, emphasizing the development of its precursors: discursive metacognition, the ability of the mind to consider its own non-sensory contents. Throughout the discussion of the evolution of animal choice and human freedom I will emphasize the dialectical nature of choice and evolution, arguing that choice becomes a factor in evolution when it emerges in non-verbal animals and starts playing a decisive evolutionary role with the appearance of humans.

Part I

FREEDOM AND ITS FORMS

Chapter 1

Basic Characterization of Freedom

One of the core problems besetting the philosophical discussion of freedom is that disagreement between different approaches seems like a war of definitions. In this way, different schools accuse their opponents of re-defining the subject so that the word 'freedom' does not make sense anymore and diverges greatly from the meaning that emerges from its common use. Then the whole discussion starts seeming futile: one might think that, since the definitions of the subject in question differ so much, it well might be that the opponents are discussing different things. In order to avoid this problem, I will start with identifying rather than defining freedom:[1] pinpointing the focus of the discussion. Here I will rely on the day-to-day use of the word 'freedom.' This approach will yield a definition that would be necessarily vague, yet it will be at the basis of any approach to the topic – for those that wish to maintain the reality of freedom as well as those that deny it; in other words, it will create what Adler calls a "topical agreement."[2] Then, I will suggest a number of characteristics required from any approach to freedom in order to make good of the common use of the term – the desiderata of freedom. Following that, I will propose several characteristics of a free entity that are needed to support the desiderata – the specifications of freedom. Then, after addressing the main problem of freedom no discussion on the subject can ignore, that of the seeming contradiction of embodied freedom, I will proceed to discuss major accounts of freedom, those that deny it and those that allow for it. Based on the analysis of the different accounts in light of the desiderata of freedom and its specs, I will argue that the compatibilist approach is the most promising and then will address the major criticisms of this approach as a challenge it has to answer.

THE COMMON MEANING OF FREEDOM

As Woody aptly notes, the concept of freedom is so frequently invoked that an inquiry into it must begin with its common meaning, or with the meaning that can be extracted from its common use.[3] Such use can roughly be divided into two types that are reflected in the philosophical debate as negative and positive freedom.

Negative freedom is freedom from impediments to agents' actions. Berlin, one of the leading proponents of the virtues of negative freedom in the twentieth century, notes that this is different from the absence of frustration – the latter can be achieved by eliminating desires.[4] Instead, negative freedom is the "absence of obstacle to possible choices and activities – absence of obstructions on roads along which a man can decide to walk."[5] This view does not ignore the inevitability of constraints – after all, our physical nature is a constraint to our desire to fly without the help of technology, and physics as of now seems to prevent time travel; these are hard to ignore. Negative freedom is a matter of degree, proportional to the available opportunities.[6]

The main line of critique that has been levied against the notion of negative freedom is that freedom needs a goal, not only absence of obstacles. Freedom is exercised by agents endowed with intentionality, and as such it is always about doing something, or at least wanting to do something; therefore, a conception of freedom that ignores the purposiveness of choice is lacking.[7] To answer this criticism, a variety of conceptions of positive freedom have been proposed before and after Berlin. For example, Rousseau in his *Social Contract* argues that freedom can be realized only through participating in the General Will of the society of individuals banded together for the purpose of self-preservation and in a submission to such society's authorities.[8] Kant distinguishes between positive and negative freedom by identifying the former with acting in accordance with the law of "pure and, as such, practical reason," with no heed given to impulses and inclinations.[9] Other conceptions of positive freedom have been offered as well.

The problem with these objections to the value of negative freedom, notes Berlin, is that the lack of a definite purpose or type of purposes does not make the removal of obstacles to agent's actions meaningless.[10] Moreover: postulating a privileged goal goes contrary to the core of freedom, i.e., having the agent making her own choices, as such choices can pertain not only to means but also to goals and ways of achieving these goals. Therefore, if we are to talk about positive freedom, we will need to abandon the approach of setting *a priori* goals for choice.

What can be the essential characteristic of freedom that follows from its common negative characterization? One does not need to labor hard to discover a common definition, one that aligns with freedom as absence of obstacles to chosen actions: "*the ability to do what we want*,"[11] choosing how to act, opting for this or that way, this or that alternative. This is the way freedom is used in the context of a teenager who wishes not to be restricted by parental supervision and in the case of an ethnic group that wants to get rid of foreign domination. Here, no privileged goals are postulated, nor do we have any preferred means. This definition is, admittedly, very vague. It does not specify the kind of willing involved: any willing qualifies, whether it be rational willing, willing determined by desires, limited or unlimited willing. The nature of ability is similarly not defined: it could be an acquired ability, a genetic endowment, or a mix between the two. Finally, the subject is not clear: the *we* of freedom might pertain to individuals or to groups. The individual might here be a rational self, however we define it, or an instinctually driven one. Would it include only human beings or other animals too? Would it apply to any human or exclude certain members of our species, e.g., nonverbal children, from the community of free agents?

However, with all its vagueness, the definition of freedom as the ability to do what we want identifies the subject well enough to have a common locus of discussion for different theories of freedom, those that argue for its existence as well as those that deny its reality. Many of the theories will answer the questions stated in the last paragraph differently and derive different conclusions from considering them, yet this initial definition of freedom enables us to continue the investigation. Hobbes emphasizes the ability to carry out this or that action and the absence of hindrance, by stating that "a Free-man is *he that in those things which by his strength and wit he is able to do is not hindered to do what he has a will to*."[12] Hegel sees freedom as tied together with will, as basic to its determination as weight is to material bodies;[13] when he later labels freedom to do as one's "natural drives" suggest as arbitrariness, he further qualifies the basic form of freedom, to do as one wishes, by qualities of the will rather than denying it.[14] Mill treats freedom as doing what one desires.[15] For Schopenhauer, freedom is "merely the absence of hindrance and restraint,"[16] yet here the absence of hindrance assumes that the agent can and wills to act as she wishes.[17] Wolf argues that free behavior is governed by the actor's will and the will – by actor's own desires.[18] Harris, with other hardcore determinists, suggests that meaningful freedom requires being conscious source of one's actions and the ability to "slip the influence of impersonal background causes,"[19] yet this is merely a clarification of what would it mean to do what one wants. The same applies to Sen's emphasis on economic and educational enablement of freedom[20] – such enablement can hope to be effective only if there is the capacity to act as one wishes. All have one common subject: freedom, identified as the area of one's ability to do as one pleases.

The vague common definition that identifies the subject of freedom has three important components that any theory of freedom needs to address: the ability, the action, and the self. These three have to be connected in a way where it is the self that has the ability to will something and to act on what it wills, as I will argue in the next section.

FREEDOM'S ASPECTS: ABILITY, ACTION, AND SELF

In the context relevant to freedom, ability has two prongs: it pertains to being capable to wish for some future state of affairs and to being capable of carrying out some action in order to achieve it. If an agent is incapable of either envisioning, in however simple way, some future and desiring for it to obtain, freedom is impossible. As the proponents of positive freedom rightly observe, agency entails intentionality. This, in turn, necessitates a mental disposition like desire, if it is to relate to something by wanting it. In order to want something that is not immediately sensed, the agent should be able to imagine it in some way – as a feeling of pleasure caused by sensory impressions, as a state of one's own mind in regard to knowledge about self or society, etc.

The ability to wish for a state of affairs, however, is insufficient for freedom – freedom is about trying to achieve the desired state, i.e., about acting in the world. Unlike merely possessing free will, in the sense of the ability to construct a desired state and wish for it to happen, freedom focuses on making it happen, as Hannah Arendt agues in the context of political freedom.[21] Indeed, the lack of ability to try and carry out the choice makes freedom meaningless. Confined to mind, it amounts to no more than a mental exercise that is not principally different from imagination: imagining a state of affairs and pairing it with desire, where the latter might well happen automatically, as in the case when an organism that can feel pleasure imagines a pleasurable state.

Yet neither the ability nor the action are possible without the self. Imagining some state of affairs that is distinct from what is the state of affairs now and acting in order to make it reality requires a locus of willing and a source of action, which must coincide if the action is to be driven by willing. This locus is essential for the "we" of freedom, its "up to usness" – it is the agent who is forming the desire and acting, not somebody else.[22] This locus is to be acting against the backdrop of circumstances, in the world that has to be changed for the desired state of affairs to be achieved. Thus, we need a clear separation between the agent and the world, the internal and the external, the self and the non-self – without that intentionally acting in the world will be impossible.[23]

A viable conception of freedom, then, should be of a self capable to set goals and act to achieve them in the world.

THE DESIDERATA OF FREEDOM

What would be required from a conception of freedom that makes good on the promise of the idea of freedom as doing as one wishes, or of a self capable to set goals and act upon them in the world? Several criteria for examining competing conceptions of freedom follow from this definition.[24] Specifically, free entity should have access to realizable alternatives, be active in the world, be endowed with imagination, capable to control itself, and be consciously intentional.

1. Access to Realizable Alternatives

Any theory of freedom should account for an agent having access to alternatives of action: if there is no way individual entities can act but one way, the "*what we want*" in "*the ability to do what we want*" loses any meaning. This is because we either cannot do what we want: we might want A, but the only way we can act is B; or the world always coincides with agent's will, e.g., because of agent's omnipotence.[25] The latter is not feasible for physical actors: due to the material nature of the world, it would be subject to laws of physics that will limit the ability of any physical agent. The former, if we consider ontological possibility,[26] looks problematic as well: even a prisoner who was born and grew up in a labor camp has the possibility to attempt an escape; this alternative is potentially conceivable. However, if we consider practical aspect, taking into consideration indoctrination, lack of education for critical thinking, conditions of hard labor, and the near-certain outcome of being killed when attempting to escape, the practical possibility of the agent seeing an escape as an option for acting is vanishing. Therefore, access to alternatives is crucial for doing what one wants, even though in practice the range of alternatives one can choose is usually more limited than the alternatives one can conceive of.

2. Imagination

To make a choice, one should be able to imagine its realization in the world. If an action originating in an agent was such that no realization of it has been imagined by the said agent, then we cannot talk about meaningful "want." Electing to do something requires imagining the desired outcome at the very least – otherwise it would not be doing *what* one wants, as the agent would not have access to the *what.* Reaching out for a bar of chocolate after imagining what would be the consequences of it is a free act, yet jerking one's hand in an involuntary movement toward the chocolate bar while thinking about the meaning of freedom cannot be seen as a free choice to grab the said snack. From here, freedom requires mentality,[27] and specifically the ability to imagine outcomes of our actions in some form.

3. Activity in the World

This is the "do" aspect of the common definition. Choice made but not realized is, to use Hegel's terms, a merely formal choice, choice that lacks effectiveness. We cannot talk about this choice being made, only conceived. Without the self being active in the world, the choice cannot be carried out. It should be noted that the character of agent's activity in the world can vary – the crucial point is to have some sort of such activity, some sort of self that can relate to the world in such a way that it can act in it.

4. Self-Control

Self-control addresses the "I" component.[28] One of the factors that differentiate free action from merely spontaneous action, i.e., an action that originates in the individual, is that the agent controls itself. There can be two types of examples of an agent that does not control itself: external control and lack of control.[29] An individual who is controlled by an external entity, e.g., through hypnosis, has no choice but to follow another's command: he does not control himself. A kleptomaniac, while being fully aware of her actions and being able to imagine the consequences clearly, still does not control her actions either: they have a degree of automaticity due to mental condition. In both cases the agents are not free since they lack self-control.

An agent that controls itself, on the other hand, in relation to its action in the world is the initiator, acting out of a wish or desire to accomplish something rather than merely an instrument at the hands of something else, be it another agent, hand of fate, or material determination. It can cause itself to move into a range of states it is capable of moving into, e.g., change location in space, quench thirst, or learn the meaning of "extrajudicial."[30] Such an agent is characterized by being *svatantra*,[31] or its own system, as Sanskrit grammarians with a philosophical bent put it. It is the initiator of action in the world with a view of certain results, out of its own wish or desire, and other entities participating in such action are, from the agent's perspective, accessories or tools. The agent has a principal position, it starts and stops the action on its own and cannot be substituted by something or somebody else.[32]

It is important to note that self-control does not exclude determinism, neither internal nor external. Moreover, determinism internal to the agent is crucial for self-control: an agent whose actions are random cannot control itself. External determinism also can be accommodated under the notion of self-control and even help it to become more efficient.[33] Forces of gravity do not preclude a fox from controlling itself: a fox's central nervous system is adept at controlling its skeleto-muscular apparatus under the conditions of gravity. If anything, it makes it easier for the fox's brain: the constancy of gravity enables automatization of certain components of movement, e.g., the calculation of the muscular effort needed to jump a certain distance. This enables focusing mental resources on other tasks. Only when no alternative courses of action are available within the constraints imposed by the world context in which the self acts, can we talk about lack of self-control.

5. Conscious Intentionality

Intentionality ties together imagination and activity in the world to guide the latter by the former. This is the connection between the "do" and the "what we want" of the definition: a targeted action, where the agent means to do something specific following imagining it being realized in the world. When controlled by the self and chosen from more than one alternative, such action can be considered free on account of being what the agent wants.

WHY NOT OTHER CRITERIA

There are several other criteria of freedom that have been suggested in the literature. The principal ones are the requirement that a free agent would be able to choose or act in a way different from the way it actually chose or acted, the condition that in the case of freedom an outcome of agent's decision should not be predictable, and the rational nature of the decision. I argue that none of these is necessary for freedom.

Why Not "Could Have Done Otherwise"

Many accounts of freedom, and specifically most modern accounts that argue for the incompatibility of freedom and physical determinism, interpret self-control of a self acting in the world, i.e., choosing between alternatives, as "*could have done otherwise*."[34] The ability to choose between alternatives that is up to the choosing self is taken as synonymous with the possibility that the agent could have chosen an alternative different from the one actually selected. If this possibility is lacking, i.e., if the agent could not have made a different choice, then we cannot talk about real freedom, as, presumably, factors other than the self determined the outcome.

The lack of the possibility to choose otherwise can be due to our physical nature: as material beings, we are subject to the laws of physics. Each and every part of our organism is physical and thus fully caused by prior events.[35] Therefore, as physical beings, we cannot but be fully determined. Alternatively, there might be no possibility to choose otherwise as a result of a particular personal history that not just predisposes the agent toward a certain choice but eliminates any other possibilities from the agent's repertoire. For example, Wolf brings up a case of a person who was raised as a racist.[36] He has never known any better, he was never exposed to another way of looking at people but the presumption of the reality of biological races and their assumed behavioral characteristics. Is this person free not to demonstrate a racist attitude, asks Wolf; her answer is negative. In a sense, he would be similar to the one who was exposed only to one language and was never told that many others exist, and thus, upon hearing a new language, might well think it is incoherent mumbling. He could not have done otherwise. And if he could not have done otherwise, there is no choice, and therefore no freedom.

Many ethicists take a major issue with this account, as they do with any deterministic approach: it is hard to justify blaming somebody for an act that was not free, that the agent could not have avoided. Therefore, those who buy into the notion of could-not-have-done-otherwise, try to develop examples where the agent could not have done otherwise yet still bears responsibility, i.e., we still can think of an agent's choice, a choice originating in the agent's aware self. Some of those accounts focus on character.[37] A person of a certain character at least sometimes acts in a way where he could not have done otherwise. Martin Luther declared that much in his "Here I stand. I can do no other." Yet one would be hard-pressed not to assign the responsibility for a stand that has been so thoroughly thought-through, resulted from much deliberation, to Luther himself. The same would, perhaps, apply to a person who had neglected his own development, indulged in binge watching of teenage TV shows instead of studying, and then failed his finals. He could not but fail the finals, yet the responsibility rests squarely with him in this case.

The character argument has a major flaw that seems insurmountable: the essentially metaphorical character of its claim. Or, to be more precise, it is guilty of considering an essentially metaphorical claim as an argument. Martin Luther's declaration cannot be seen as something he could not have avoided by choosing a different route on April 16, 1521, and not coming to Worms altogether – we have little reason to believe all other roads were blocked. It was his decision to follow his own convictions, even if the convictions he's formed in the preceding years all supported his decision to proceed. Moreover: the development of such convictions can hardly be seen as something but his choice. If, on the other hand, we deny the latter, then we deny his freedom altogether. Same applies to the lazy student. The linguistic use of "I could not have done otherwise" is very different in the case of Luther and in the case of somebody referring to her digestion. In fact, in this case few speakers would say "I couldn't but digest this sandwich" – precisely because the use of "I could not have but" is usually metaphoric. Even in the case of action under duress it denotes the unwillingness to suffer extreme harm rather that the ontological impossibility of choosing an alternative course of action.

Another set of examples that is supposed to demonstrate situations where an agent can be assigned responsibility since he acted freely yet had no choice but to act in a certain way has been provided by Frankfurt[38] and since then expanded by others. Frankfurt's examples outline circumstances that leave only one alternative yet do not move the agent toward specific action. Specifically, he brings up cases where a person does something on her own will yet she would have done the same thing if she were to decide differently. For example, one can imagine a team of government agents who are interested in luring a suspect into a vehicle disguised as a cab. They would much prefer that he enters the cab on his own will but are ready to pounce and push him in if needed. The suspect does enter the cab without any extra action from their side, yet he really had no choice: this or that way he would be inside.[39] Frankfurt constructs similar cases with action and potential coercion that comes into play only when the individual in question does not act as the coercers desire. In all those cases, "what action he performs is not up to him," concludes Frankfurt.[40] This, however, is incorrect. We are dealing with two separate cases here: one of free choice and one of coercion. If the individual in question chose to perform the act on his own, it was his free choice. He did have other alternatives prior to making the decision to act in a certain way and at the point of acting: there were at least two alternatives he could possibly imagine, making a move or not making it. It is past the decision and the initial move to act when the lack of real alternative is discovered, and even that only if the agent decided freely to act in a certain way, e.g., not to get into the cab. The lack of viability of one of the alternatives becomes apparent to the agent *post actum*. The illusion of lack of choice in these cases results from defining action in such a way that it stretches over from the point of decision to the perception by an interested other, whether the decision point and the move to act is what is important for determining whether the decision was free in a meaningful sense.

The criterion of *could have done otherwise* therefore has two different meanings. One is essentially metaphorical and denotes the condition where the agent could have done otherwise. The other is a result of confusing free decision and move to action by the agent and actions by other agents that come after this point. Therefore, it should not be included in the desiderata of freedom.

Yet what can we do with Wolf's racist? The problem with this case is that he could decide otherwise and review his convictions. Being endowed with language and therefore the capacity to address his own thoughts, work out new criteria for evaluating states of the world and own motivations, communicate with others and change personal convictions based on argument and evidence, he can mend his ways.[41] In fact, much of human history is a tale of a variety of people in different epochs doing just that, from Akhenaton deciding to promote quasi-monotheism despite intensive indoctrination to the contrary and never being presented such an option, to those who were raised racist yet then changed themselves, Wolf's exact case. Similarly, a person who has used one language and engaged in communications with others can easily comprehend a situation where a different set of vocal signs is used – after all, he himself had been in a situation of language learning.

The problem that remains is the one raised by hard determinists, that of our physical nature, combined with the environmental context, dictating exactly the course of action, thus making free choice a fiction. This problem will be dealt with later, in the section discussing determinism as one of incompatibilist accounts of freedom.

Why Not "Could Not Be Predicted"

Another criterion that is frequently proposed as distinguishing between freedom and the lack thereof is predictability. The argument here is that if we can, based on certain events we are capable of observing, reliably predict individual's choices, then these choices follow from what we observe.[42] If such observed events are not subject to conscious control, or not under individual's control at all, then her choices cannot be considered free. For example, if we can accurately predict behavior based on person's background – individual history, genetic makeup, employment, or anything else which does not result from conscious choices made by the individual in question – then this is no individual's decision in any meaningful sense. If we can predict, with high accuracy, individual's behavior based on measuring his brain's electrical activity prior to the decision being made, then these are not his decisions, not his choices – he has no control over those brain occurrences of which he is unaware.[43]

Long before the advent of neural measurement this issue was raised in relation to divine omniscience.[44] If God knows everything, including the future, and this knowledge cannot be mistaken, it seems to follow that nothing can happen differently from this one way known to God. While this problem is as old as the belief that the divine is all-knowing, there is its more modern formulation that is relevant to foreknowledge that is not divine:

(1) Yesterday an infallible belief regarding what agent A would do was formed, belief that B.
(2) It is necessary that yesterday belief that B was held, as it was formed yesterday.
(3) Necessarily, if there was an *infallible* belief that B yesterday, then B.

(4) Therefore, it is now necessary that B.
(5) If it is necessary that B, then it cannot be decided and done otherwise, i.e., it is not in the power of agent A to decide and do otherwise.[45]

The theological aspect of this problem has been addressed by many religious scholars and philosophers: Saint Augustine, Boethius, Thomas Aquinas, Maimonides, Gersonides, William of Ockham, Leibnitz, and others. Most of them were interested in clarifying the nature of divine omniscience, the meaning of the infallible foreknowledge that is characteristic of omniscience. For example, one can argue, with Boethius, that divine foreknowledge exists outside of time, and thus knowing X before it happens fails to lead to X being determined.[46] Maimonides, on the other hand, takes a different approach. In *Eight Chapters* he recognizes the problem, further described from different angles in *The Guide for the Perplexed*.[47] If we are to acknowledge God's omniscience, we must admit that he knows what it is to happen in the future. Thus, it might seem that the future is fixed – otherwise how would it be known? Therefore, there is no free choice if we are to assume divine foreknowledge. However, notes Maimonides, this would be the case only if we assume that divine foreknowledge is similar in character to human knowledge. Yet this assumption is not warranted – we cannot know the nature of the divine foreknowledge.[48] Therefore, we cannot argue that it is incompatible with human free choice.

This line of thinking is applicable to the problem of prediction even if it is not expected to involve omniscience. Predicting an event consists in correlating certain observations made in the past, *p* and *q*, drawing a conclusion about causality, i.e., *p* causing *q*, and then, when observing *p*, predicting *q*. Yet from this it cannot be concluded that "one event is somehow in the power of another."[49] Predictability by itself, as Ayer notes, does not imply constraint;[50] nor does it imply necessity and the lack of free choice by the agent, just like correlation does not imply causation – these are conceptually different. In order to progress from prediction to an argument for causality, there is a need to develop a convincing theory that will demonstrate that agent's choice cannot be the cause of agent's action, and this goes beyond mere predictability. After all, one can reliably predict person's IQ based on her shoe size, until the age of ten or so, yet the size of one's foot does not determine intelligence. Therefore, "could not be predicted" should not be included in the desiderata of freedom.[51]

Why Not "Rationally Decided"

Traditionally, freedom is related to rational reasoning, the ability to weigh options in light of general criteria applicable in different situations.[52] In the case of freedom, this would be commonly assumed to require that the agent considers the situation of choice as a situation where a way of acting should be selected and followed, behavioral alternatives – as different options from which it needs to select one to follow, and agent's own actions – as something that needs to be done in order to carry out the choices. Otherwise, argue the proponents of this position, the choice is no better than random movement or uncontrolled twitching.

At the core of this view lies a specific way to understand self-control. It assumes that an agent cannot control itself but rationally, considering itself as a choosing agent that needs to make a decision, and thinking about itself making choices, thus employing reflexive cognition. This way, self-control will necessarily require rational deliberation, where the agent is aware of its own deliberating capacity and is employing it consciously.

The problem with this view is that self-control can be achieved and choices can be made without rational reflexivity. The view that equates rational reasoning with choice fails to distinguish between making choices and a particular way of choosing that is praised in philosophy and considered to be the best option for humans, i.e., rational choice. While there are different accounts of rational decision making, all of them include deliberate consideration of choice options according to certain principles that are commensurate with goals, where the agents consciously evaluates the merits of the available options in accordance with some overarching criteria.[53] Yet choosing between available options can be achieved without rational deliberation. This would be the case when different options are simply manifest to the choosing agent and associated with sensory images stored in memory, and the agent is such that it can exercise preference toward this or that feeling, or has interests.[54] At a minimum, such agent should be capable of registering what is immediately given through the senses and of storing the sensed images in memory for future retrieval – without that it would not be possible to have behavioral options to choose from. Furthermore, it will have to be able to associate images, to generalize over different sensed and remembered mental contents – otherwise comparing between options would not be possible. Of course, the ability to distinguish between stored and intuited images is necessary as well. No effective choice can be made when the agent does not apprehend the difference between the food it remembers and the food is it encountering at this moment, for example. All this makes choices manifest to the organism and associated with feelings, e.g., those of pleasure and pain – and then the choice can be done following the preference toward pleasure, for example.[55] Having control over its material structure, e.g., its body, enables the agent to carry out its choice, or at least attempt to do so. Arguably, a person who, feeling fearful, makes a momentary decision to fight or to flee, has little, if any, time for deliberation, yet the decision is still a choice; in another similar situation a different choice can be made, and another agent can make a different choice as well. In less threatening circumstances, decisions made regarding what meal to buy, what shoes to wear, or how to express this or that thought are frequently made without rational deliberation, yet it is hard not to see them as choices without begging the question, or redefining choice as necessarily rational.

THE SPECIFICATIONS OF FREEDOM

What would be required of a free agent in order to realize the desiderata of freedom? In other words, what would be the capacities an entity should possess in order to have a chance to be counted as free and what kind of environment is necessary for such an entity to operate freely? To answer these questions, the free self and the free self's world need to be addressed.

The Free Self

A free entity has to be a self, not merely something separate from other objects in the eyes of external observers but an objectively describable locus of activity, where the source of such activity is within the entity itself. To be a self in the sense relevant to the question of freedom is to be a source of determinism, its subject, and not merely its recipient, or its object. Without such self, there would be no agent that can make choices: making choices means causing change.

The free self also cannot be merely a theoretical entity postulated for reasons as good as they might be. Dennett describes the self as playing a role that is conceptually similar to the center of gravity – nicely defined, well delineated, and well-behaved *abstractum*, fictional object, that plays an important explanatory role in the philosophy of mind, similarly to the role the concept of the center of gravity plays in physics.[56] Such a self would be a theoretical construct that will have only those features that the theory endowed it with, as it does not refer to anything existing in empirically reality. Yet this sort of self would have no efficacy in the world: theoretical constructs cannot act in the material universe, neither can they will. Their existence is confined to theoretician's mind, similarly in terms of their lack of the ability to act in the world to the way the existence of fictional literary characters is confined to the imaginary universe created by writers and enjoyed by readers. It is for this reason that such a self would never be a choosing subject: none of the desiderata of freedom can be accommodated by a fictional object.[57]

Thus, the self of a free entity must refer to an observable entity causally active in the world. To continue Dennett's line of physical metaphors, such a self should disturb the web of causality in a way somewhat similar to the impact made by a lump of matter in the fabric of spacetime. If we are talking of a material world, the term 'self' has to refer to a material entity that is not only an object of causality but also its source. Only in this way the self will be able to act in the world, as required by #5 of the desiderata of freedom (*Activity in the world*).

Before making a choice, the self has to have access to the alternatives: being able to perceive the options form which to choose (#1 *Access to realizable alternatives*) and to envision possible outcomes (#2 *Imagination*). Without the former, the possibility of choice would never arise: if there are fewer than two courses of action accessible to self, there is nothing to choose from. Yet the latter is crucial as well: if the self has no way to imagine the outcome of choice, it cannot be meaningfully conceived as making choices.

To carry out a choice, self has to control itself (#4 *Self-control*) – without it the choices would not be of this self, the self will not be choosing. As material, it has to have mechanisms that enable its body to control the activities of their parts, receive feedback from them, and act as one unit – one choosing unit. In other words, such a self will be a locus of self-control,[58] a lump of matter that controls itself. Self-control, at least the kind of self-control required for choosing, however, entails self-awareness: if an entity is not aware of its own state in reference to the material context of choosing, it cannot really choose, since it would not be able to direct and correct, if needed, its advance toward its goal. If the acting subject is not aware of its own urges, it will not be able to act upon its urge: a goat that is not aware of its hunger will not try to assuage it by advancing toward a patch of grass. If the subject of action is not aware if its own body, it will not be able to move toward the desired goal – a protist that moves with a flow of water is not moving itself, it is being moved, similarly to a speck of dust. If the subject is not aware of the location of its body in space in relation to the desired goal, it would not be able to move towards it and to correct the course if, say, the goal moves or if it encounters an obstacle. This awareness can be instantiated through different mechanisms, yet it is essential for freedom.

So far, the self can be summarized as self-determining: a choosing entity which, being aware of its own action, the options it has, and what it imagines to be possible outcomes, does something to effect its choice. Yet without intending to act in this or that way consciously there would be no self-determined choice. A robot can be a source of action which is triggered by certain environmental stimuli to which the robot had been programmed to respond. The said robot can be also programmed to receive feedback from sensors that measure its own position and the position of its target, e.g., nuts and bolts, compare the two, and adjust its own position based on the results if such comparison.[59] However, the robot is not free, as it does not author its intentions:

they are an outcome of the program that has been implanted in it. If our robot is more sophisticated and has a learning mechanism, e.g., a neural network,[60] it will not change the fundamental fact that the robot is not an author of its intentions, did not choose its goals, yet merely adjusted its workings in light of an externally defined goals. Thus, true self-determination requires conscious intentionality (#5), a choosing entity's authorship of its choices.

Therefore, realization of the desiderata of freedom requires an entity that would be self-determined, capable of envisioning possible outcome of its activity in the world, and endowed with consciousness.

The Free Self's World

For freedom to become actual, the world in which the free self functions needs to be commensurate with it.[61] Such a world should mirror the desiderata of freedom by enabling the self's realization of them. It needs to provide alternatives for the self's choice – otherwise the self will have nothing to choose from. It should be of such character that the self can act in it. It should provide material for the self's imagination. Only in such world can an entity that controls itself exercise its conscious intentionality.

More specifically, the world in which a self functions should allow it to perceive enough of its aspects to build an *Umwelt*, a self's phenomenal world, a slice of the world accessible to the self and meaningful for it, both as providing information and allowing for actions.[62] Elements of such a world should be available for the self's systems to perceive and act upon, and they should be consistent with each other and the self's abilities to perceive and to act. This way, the *Umwelt* has a dual nature: its elements objectively exist and are external to the self, yet which elements are conducive to the agent's choice is as much a function of the agent and its structure. The world in which humans and bats live is one and the same world, yet their *Umwelts* are very different. A bat's *Umwelt* has ultrasonic mapping thanks to its perceptual system, while the one of humans lacks it. The human *Umwelt* has oceans to cross, while a bat's is devoid of this feature – bats have no ability to navigate large expenses of water. The human *Umwelt* has jobs to apply to, books to read, and governments to depose. It is much more pliable than a bat's due to human abilities to intentionally effect change in the world. It is the *Umwelt* through which the world acts as a playground for freedom.

In terms of alternatives, the world should provide ways for the self to act in it. For a self desiring to consume nutrients, for example, the world might make available different foodstuffs and ways to access them. These can be plankton floating in the water, perceivable to the self capable of swimming and filtering plankton-rich water through its teeth; a patch of grass, visible and accessible to the sense of smell, and a terrain the self can navigate; a grocery store with apples and oranges to buy and a way to earn money to buy them. The availability of options to choose from is directly proportional to the degree of freedom: the fewer alternatives the agent has, the narrower is its freedom as it can be exercised in practice; these are functions of the self's *Umwelt*. In a situation where there is but one alternative, freedom is meaningless.[63]

The examples above emphasize the commensurability of the world with the self that is necessary for freedom, commensurability in aspects accessible for such a self, its physical and mental abilities. Grass is useless for a whale, and plankton is useless for a mountain goat; horses cannot read, and people lack the keen smell of dogs. If self is a light whose awareness illuminates the world of objects perceptible to it,[64] there should be entities in the world in which the self operates that are responsive to this light – otherwise the self cannot be free.

For selves whose activity in the world is such that they can only rely on readily available alternatives, incapable to change the world in such a way as to create their own, lack of ready-made alternatives would lead to inability to exercise freedom. A mountain goat without mates available in the vicinity, which it can traverse by recollection of sensory data, e.g., visual and olfactory images, will not be able to exercise freedom of choosing a mate. A human being, on the other hand, can create a dating website. For beings endowed with creative cognition that can intentionally think about the way to change the world to suit their needs, imagine the world being different, and work to achieve this state of affairs, the world is dynamic and more malleable; for being with no such capacity, the requirements for the world in which it can exercise freedom are more stringent and are directly linked to its perceptual and physical constitution.[65]

This, however, is a macro level: we as humanity can draw maps, build sail ships, and travel to other continents in search of black pepper. As individuals, our choices are much more limited for the obvious reason of our own restricted range of ability: financial, social, and intellectual. *Umwelt* is a result not only of biological species' makeup and geology but also of individual circumstances, social environment and personal history. Here, human society plays a great role in expanding or limiting freedom, even if it does not impose explicit constraints on its exercise, but this is outside the scope of this book.[66] To our purposes here, though, the world in which an agent lives, in order to allow for its freedom, should be conducive to developing its capacities relevant to choice if such capacities require development. The range of possible choices is directly proportional not only to the number of available alternatives offered by the world but to the degree to which capacities of a choosing self are developed. A goshawk is endowed with impressive killing instincts and does not require much training.[67] A lion cub who is not trained in hunting by its parents, on the other hand, will show poor performance. A human being who has never been exposed to language, a capacity definitive of human choice, will never develop linguistic proficiency. As a result, such human being will be severely limited in his ability to exercise freedom, would not be able to conceive of choices that are not associated with the sensory aspects of encountered and remembered images, as language is necessary for non-sensory mental content. It is language that allows carrying out the capacity for discursive metacognition,[68] considering not only images but the relations between images in light of criteria, as well as the criteria themselves: language is a vehicle which enables us to name different types of mental contents, categorize them, and evaluate them.[69]

THE PROBLEM OF FREEDOM

This book is not concerned with the question of whether free will is real but rather with developing a conception of freedom that would answer the challenge posed by freedom in the world and specifically with describing freedom in non-verbal and verbal animals. However, it makes sense to address the core issue of free will in order to prepare the ground for developing a feasible conception of freedom.

The main problem with conceiving of human freedom consists in the seeming contradiction between determinism and free will or, to be more precise, between the way determinism is usually understood and the common conception of free will. The common meaning of determinism here is that all events are caused by events that took place prior to them, and none can escape this causality. Specifically, material events, those pertaining to the physical, spatio-temporal aspect of the world, are fully determined by other physical events that preceded them. This view is not new, it has been a paradigm of the modern science since its inception. Moreover, as Woody notes, "it is an *axiom* that makes scientific inquiry *possible*"[70] – not the result of its theorizing or experimentation but its very foundation. Not only scientific inquiry but the whole of rational thought labors within a deterministic framework: admitting that something can take place without any cause would be admitting that something happened without a reason. This violates one of the most cherished and, perhaps, obvious principles of the rationalist approach, Leibnitz's *principle of sufficient reason* (PSR): it postulates that "no fact can be real or existing and no statement true unless it has a sufficient reason why it should be thus and not otherwise."[71] No statement, that is, except the PSR itself, which is taken as axiomatic. In the practice of empirical science, formal sciences, philosophy, law, etc., this principle rules supreme: this book would never be treated seriously unless it provides adequate arguments for its conclusions.

Yet if there are causes at work everywhere, then not only every movement of my hand is caused by something, but also every electrical impulse in my brain should be traceable to its cause. Even without getting into the details of neurophysiology, one can argue that a person who had enough character strength cannot be considered an author of her decisions: there was this strength of character first place, caused, in turn, by something else: nature, nurture, etc.[72] In these circumstances, self-control becomes inconceivable.

Physical discoveries at the level of elementary particles during the first half of the twentieth century led to Heisenberg's uncertainty principle and gave hope to some opponents of determinism: the principle provided for a measure of uncertainty that looked like a part of observable nature, thus breaking the causal chain of necessity by postulating events that cannot be clearly predicted based on prior events, and thus could be conceived of as lacking deterministic causality. However, the development of quantum mechanics restored the measure of determinism, where particle's position and velocity can be represented by a wave function and predicted with a degree of certainty.[73] In any case, such effects are not observed at the level of organic molecules and living organisms that can be reasonably suspected of making decisions – those still can be described by the laws of Newtonian physics. Yet even if we are to introduce quantum or other type of genuine randomness into the physical world and thus dethrone determinism, it will not solve the problem of freedom.

Indeterminism, or the lack of determinism, is reducible to chance, the unpredictability associated with it owes to the randomness that stands at its basis. Thus, if something is undetermined, random, then it is incompatible with agent's free will: being endowed with freedom is to author one's choices, yet chance has no authors. Therefore, indeterminism is incompatible with free will.[74] Hence, we can formulate the problem of free will this way:

(1) Either determinism is true, or indeterminism is true;
(2) Human beings have free will;
(3) Free will is incompatible with either determinism or indeterminism.[75]

Free will is commonly understood as the ability of the agent to be the *sole* author of her decisions. As Harris puts it in his popular account,

> Our moral intuitions and sense of personal agency are anchored to a felt sense that we are the *conscious source* of our thoughts and actions. When deciding [. . .], we do not feel compelled by prior events over which we have no control. The freedom we presume for ourselves and readily attribute to others is felt to slip the influence of impersonal background causes.[76]

Many admit today that there would be a variety of constraints on such free will, e.g., the case where the alternatives of choice were supplied to us by the way we were raised and the culture to which we belong; that the very way we think about things is shaped by our social environment; etc. Yet free will, as commonly understood, still allows for a decision space, however small, that is absolutely agent's own.

With these considerations and having the common understanding of free will in mind, it is clear why *(3) Free will is incompatible with either determinism or indeterminism* will follow: neither fully determined nor randomly behaving beings cannot be authors of their decisions.

Different accounts of freedom, as well as those that argue for the lack thereof, are all trying to solve this problem. Three of those are paradigmatic, even though they give rise to multiple versions: libertarianism, determinism, and compatibilism.

ACCOUNTS OF FREEDOM

Incompatibilist Accounts

Incompatibilist accounts argue that free will and determinism are mutually exclusive.

Within the determinist framework, as it was noted earlier, for two states of the world, S_1 and S_2, the earlier state S_1 in conjunction with the laws of physics[77] entails S_2,[78] or fully determines it.[79] This view has great merits. Conclusions drawn with the help of this assumption have been withstanding the test of observation and producing correct prediction since the dawn of humanity; scientific theories developed with this assumption in mind have been supported by plethora of empirical evidence; theories that have been falsified gave way the new ones, while the assumption of the deterministic material world stood firm. The deterministic view complies with logical constraints produced by reasoning as well. A reason to abandon deterministic view of the material world should be much stronger than a desire to maintain a cherished conception of action.

It is important to note that the type of determination relevant to the question of freedom in the world, freedom of agents that are active in a material universe, is empirically observable rather than logical. In other words, we are considering here contingent states of the world existing in space and time rather than logical necessities that are oblivious to spatio-temporal constraints and are not subject to change. The latter are irrelevant to the question of the possibility of freedom, since freedom focuses on choice, change in the world that is up to the agent. It is not up to any agent to make the sum of the squares of the legs equals to the cube of the hypothenuse rather than to its square, yet, if the agent is free, it is up to her to proclaim that.

Usually determinism is considered to be equivalent to the thesis that "at any time [. . .] the universe has exactly one physically possible future."[80] This way, any constellation of causes will have only one outcome. This reading of determinism seemingly clashes with certain elements of quantum mechanics, as it has been noted previously. However, this interpretation of the deterministic position seems misguided, at least in the context of freedom. If we introduce a measure of uncertainty that can be described by a formula and empirically verified, it would not alter the fundamental nature of determinism: full explanation of state S_2 at time T_2 by state S_1 at time T_1 that precede it. Even if we allow for Schrödinger's cat to be either alive or dead, its existence or the lack thereof is still fully determined by the state of the physical system that precedes the observation. Therefore, probabilistic calculations do not change the main tenet of determinism: fixing each event, each state of the world, by its previous state.

Incompatibilist philosophers argue that free will is incompatible with determinism. If we conceive of freedom as the ability to decide and act this or that way, and this decision is up to the acting agent, then the agent *can* choose this or that option, and this choice is up to the agent.[81] Yet in this case we have a problem. A choosing agent cannot change either the past or the laws of physics. Since every state of the world, including the one that encompasses agent's actions, is fully determined by the past states and the laws of physics, the agent cannot make a real choice, to act in this *or* that way.[82] It is this analysis of the concepts of 'can,' 'law,' and 'state of the world' leads us to the conclusion that free will is incompatible with determinism.[83]

There are two types of approaches to free will that embrace incompatibilism: libertarianism and determinism.

Libertarian Freedom

Libertarianism is the classical and, perhaps, the most commonsensical account of freedom. At the core of the libertarian approach is the claim that we possess a measure of absolute freedom – freedom to choose that is not constrained by anything. To put it in terms of cause and effect, there are at least some, if not great many instances where our decision cannot be traced to prior causes, it is a cause without being an effect of another cause, our individual *fiat* in the web of events, other part of which might or might not be deterministic.

One can argue that freedom cannot be unlimited: the choices are not entirely up to us. Besides obvious environmental constraints that are expressed in the alternatives to which agents have access at any point in time, there is also the constraint of prior choices: our choices limit the range of future choices, e.g., in situations where choosing to sleep longer makes the choice of coming on time impossible. This has been referred to as a garden of forking paths.[84] However, argue libertarians, it is not the number of alternatives but the choice itself that is unconstrained, entirely up to the agent: "the ultimate sources of our actions lie in us and not outside us in factors beyond our control."[85] It well might be the case that not all our choices are unconstrained but only some of them, for example, actions that have great impact on our character which, in turn, underlies our choices for years to come, actions that Kane deems "self-forming actions" (SFA); it is thanks to them that we bear ultimate responsibility for our choices, even if some merely follow from the characters we came to possess because of the SFAs.[86] In any case, libertarian freedom requires an element of unconstrained choice. It entails that the choice is open: the agent could have chosen a different option given exactly the same past – the same past of the universe, that is, including agent's own background and history.[87] This means that there can be different futures given the same past.[88]

This kind of freedom is incompatible with determinism: if my decision is caused by something else, whether I am aware of it or not, it was not mine, was not really up to me, and hence was not free;[89] there can be no more than one future for the same past, or, if we are to take quantum mechanics into consideration, the choice between these futures cannot be free but only random.[90] The agent-centric aspect of choice is crucial for making sense of libertarian freedom, as it is necessary for giving this account a chance to answer the desiderata of freedom specified earlier. Unpredictability alone would not do: if the mechanism behind the choice produces random outcome, then the outcome is not up to the deciding agent, and the agent is therefore not free.[91] No matter if my choice was caused by urges that the agent does not control, by her upbringing or by her genetic makeup, if the agent was not the one to decide, with equal possibility[92] to go for this or that alternatives, the choice would not be free for a libertarian.

The main issue with the libertarian conception of freedom is the *intelligibility problem.*[93] Given what we know about the world and the place of physical agents within it, unconstrained libertarian choice appears to be impossible. Individual *fiat* does not have a place in the deterministic picture of material world, where everything has its cause. In such a world, agent's decision will have a cause just like any other event, and the relationship of causality will not be principally different in this case from the mechanical interaction of billiard balls.

One way to salvage the libertarian conception of freedom is to postulate some type of extra-empirical relations of causation, with an uncaused cause outside space and time. Kane calls these "'extra factor' strategies":[94] noumenal self, a type of causation not familiar to the sciences, non-material self or soul, and transempirical power centers are some examples of this approach. Yet these strategies would be vulnerable to the charge of obscure arbitrariness: proposing a theoretical crutch to save an idea toward which we have a strong positive prejudice is a poor way of philosophizing. Kant's suggestion that we need freedom to make sense of morality and responsibility, that we have to presuppose "an inscrutable faculty which no experience could prove, [. . .], faculty of freedom"[95] that makes moral law possible, is similar in terms of its way of arguing, and thus is as vulnerable to similar critique. Thus, to cope with the intelligibility problem, libertarians must show how their conception of freedom-through-indeterminism, the up-to-the-agent nature of choosing that requires a disconnect between agent's choice and prior events independent of agent's choices that have causal powers toward future events, is intelligible.

One such account is provided by Kane and pertains to the self-forming actions (SFAs) mentioned earlier.[96] In moment of great quandary decision makers face competing visions of what they should do. The stakes are high, the options are salient, each has strong pros and cons on agent's terms. These can be moral dilemmas, as in Kane's oft-used example where a person who is torn between the option of helping another human being in need and being late to a decisive meeting of her career, and the alternative of making it to the meeting and ignoring the suffering of another. These can also be situations of fight or flight, where both strong incentives and well-comprehended risks exist for all accessible alternatives. In those cases, suggests Kane, relevant regions of our brains move "away from thermodynamic equilibrium [and lead to] 'stirring up a chaos' in the brain that makes it sensitive to micro-indeterminacies at the neuronal level."[97] This, according to Kane, "temporarily screens off complete determination by influences of the past."[98] Here the outcome is undetermined yet can be willed by the agent either way, and thus it is voluntary: there are concerns of which the agent is aware that relate to each choice alternative. When a certain alternative wins, we make it prevail by deciding, by making one set of concerns prevail over the other.[99]

Why the choice here would not be a random outcome? Because, argues Kane, it is *willed* by the agent, and willed based on *reasons* of which the agent was aware. After all, whether it is fight or flight, the deciding agent had concerns regarding each alternative, with such concerns thought-through, cursorily noted, or vividly felt. Of course, before the choice outcome occurs, the agent does not control it: this a necessary feature of the proposed solution, the indeterminacy of the outcome, and a central feature of libertarianism. However, writes Kane, regarding the alternative outcomes "it does not follow [from here] that [the agent] does not control or determine which of them occurs, *when* it occurs."[100] The choice here is undetermined, yet, for Kane, it does not mean that it is uncaused – it is caused by the agent's will and by effort the agent invested in comparing the alternatives and making the choice.[101] The indeterminacy on this account is a hindrance to each of the alternatives rather than a factor in deciding, something the muddles agent's goal-directed activities and akin to static noise that has to be overcome by meaningful signals.[102] Agent's decision means having one alternative overcome the indeterminacy.

The main issue with this account is that it does not establish free will in any meaningful sense. It is unclear how the decision itself – not the buildup toward the decision, the sets of competing cognitions, but rather the choice – is made by the agent. If we are to stick with Kane's account, then free will remains a mystery, to use van Inwagen's phrase:[103] it is unclear how the will, which is not traceable to prior conditions and remains somehow above the fray of overlapping and competing neural networks, intervenes in the randomness of electrochemical chaos and makes a choice. In order to fix this problem, alternative accounts have been suggested. Balaguer,[104] for example, proposes an important clarification of Kane's scheme. When the decision is torn yet one option is still being selected, no consideration is known to the choosing agent that would point to a specific alternative – otherwise it can be claimed to be determined. This is an arbitrary choice, yet as such it is the sought after free choice.[105] What makes this choice free, then? For Balaguer and for Kane, it is the consciousness, the intentionality, and thus the responsibility the agent has for the choice.

This solution, as honest as it is to the spirit of libertarianism, seems flawed, because it makes decisions arbitrary. A plausible explanation for the experience of choice on this account might well be that the scale on which the alternatives of action are placed is tipped by a genuinely random neural event, e.g., of a quantum origin – Kane himself makes references to quantum events;[106] this idea of random sources for human decisions is quite old: Lucretius suggested that free will is traceable to the random swerves of atoms.[107] This way, libertarian conception of freedom, even when limited to certain type of actions, becomes unsustainable.[108]

Determinism

Determinism as a philosophical approach denies the existence of freedom because it finds it irreconcilable with determinism which the proponents of this approach see as descriptive of nature. Since Nature is deterministic and any possible candidate for free agent is part of Nature, freedom does not exist.

The roots of modern deterministic arguments are in the common interpretation of Spinoza's position on free will. In his *Ethics*, Spinoza states in proposition 28 that an individual thing, which is necessarily finite, "cannot exist or be determined to act unless it be determined to exist and to act by another cause which is also finite and has a determinate existence."[109] The proof of this proposition relies on previous argument for all existence being in God/ Nature[110] yet finite entities not being able to follow from one of God's infinite attributes: otherwise they would be infinite. To our purposes, we can see it as the impossibility of deriving a physical law that would not have universal import from a physical law that would have a universal import: any formula for calculating gravity will have to obey by the laws of gravity and will apply to all possible material objects. However, if we are to consider any individual physical entity, limited in space and time, it would always be under physical laws, since it is in Nature, yet determined to action by other individual entities that are also under physical laws. Therefore, all finite beings and their acts would be caused by other finite beings and their acts, in accordance with the laws of Nature.[111] No causality the sole source of which is in the agent can find its way into this scheme, as long as the agent is material.

It is this feature of the candidates for the role of free agent that makes the crux of the deterministic argument. If we consider humans, it is hard not to note that humans are physical. Our thinking, even the most abstract one, is supported by the brain: stopping blood flow to the brain makes human individual to stop demonstrating any detectable signs of decision making. The only ways we have to carry out our decisions are our physical organs, be it tongue or hands, which are physically connected to our brains through the nervous system: sever this connection, and decisions will lose efficacy. The necessity to hook up sophisticated electronic equipment to the brain or other elements of the nervous system in order to assist people with motoric impairments further highlights our physicality: we can map specific neurochemical pathways and discern signals that instruct our organs to move. Memory, a function that plays crucial role in choice as it enables the comparison of alternatives, is physical as well: recent experiments point to the possibility of transferring learned behaviors by implanting biological matter from a trained animal into a naïve one.[112]

The cognitive apparatus that supports choosing, as complex as it might be, is built from the same components of which the rest of the universe is composed: molecules of matter, with the neural system relying on electrochemical exchanges. It is physical because it is built from these components, not for any other reason: if it were built of anything else, it would not be finite. Atoms, as we know from natural science, are deterministic in the relevant sense: observing their workings and studying their structure leads us to this belief. Spinoza's argument makes sense only if we rely on this observation – otherwise it is impossible to establish the relevant physicality and thus finality of bodies. This does not leave any space for free will: every event is determined by prior events. In many cases, we now can explain how the process of such determination works or simulate it in the lab. At the macro level, we might conceive of free decision making based on observing individuals acting, yet only if we ignore the physical nature of the individual. Such physical nature is built upon the level of components, and it is the examination of this micro level that establishes causal necessity. This is similar to the way that while looking at a bridge creates an illusion of a whole, in reality the bridge is composed of atoms of matter, and its features are determined by their features.

Standard interpretation of quantum mechanics suggests that the world is not deterministic in the way the laws of Newtonian physics led us to believe. However, this does not salvage the notion of incompatibilist free will, as I argued when criticizing the libertarian account in the previous section. If it is the randomness element that plays a role in which alternative the supposedly free agent chooses, then it is not a free choice in any meaningful sense: randomness is as much an enemy of free will as causal necessity. Moreover, a degree of randomness does not contradict deterministic lawfulness, only introduces a probabilistic element into it. This or that way, freedom that has a chance of answering the desiderata specified earlier cannot exist in a physical world, per determinists' argument.

Perhaps the most common example of the inefficacy of what is consciously perceived as free choice is determinists' interpretation of Libet's experiments.[113] In his research, Libet discovered that brain's electric activity "wells up" before certain decisions are made, i.e., reported by the experimental subject as made; the profile of activity was found to be linked to particular decisions. For example, a subject reports that he is going to move a finger at time t_1, yet the electric brain potential linked to a finger movement, what Libet and his colleagues deemed *readiness potential*, appeared already at time t_0 that is earlier than t_1 – significantly earlier than it is required to report the decision. Later findings using more sophisticated measuring equipment detected readiness potential seconds before the decision was reported by the subject.[114] Determinists interpret that as a proof that the decision the subject makes merely reflects the determination of the organism to act in a specific way, a determination that becomes available to consciousness after it has occurred. These findings have been commonly interpreted as a proof of the illusory, epiphenomenal nature of free will – an account that argues that all mental events are caused by physical events, yet no physical events can be caused by mental events.[115]

Why then does the intuition of free will, an intuition that shapes our moral thinking and, perhaps, much of human interaction at all levels, exist at all? Some determinists would argue that our conception of freedom is no more than an illusion, an honest mistake of the mind that might not be much different in its nature from optical illusions. Similarly to how the latter are an unfortunate outcome of our otherwise quite impressive perceptual system, the illusion of freedom is a byproduct of the way our body, and particularly the nervous system, works.

This account, however, does not explain why the illusion is so strong and consequential to our lives: after all, illusions rooted in our genetic makeup[116] that are simply not harmful usually play little role, and those that are harmful are weeded out by evolution since they are detrimental to survival. This can be explained by suggesting that the illusion of freedom might also coexist, lead to, and be reinforced by the delusion of freedom. The delusion here would be an odd thought that claims to describe reality yet is not reflective of what human individuals hold following their own life experiences or socialization.[117] In its extreme form, it can be a paranoiac's conviction that his neighbors are plotting to murder him. In a more common and, most regrettably, socially acceptable form these can take shape of ethnic stereotypes, at times no less fantastic than a paranoiac's fear of neighbor's lawn mower that in his delusional mind is an advanced spying device. The delusion of free will serves an important social role, lays the foundations of justice system on all levels, etc. – just like the delusion that nations are somehow natural rather than constructed entities serves a formative role in domestic politics in many countries. As it plays such an important role, the society – again, at all levels – fosters the delusion of freedom, socializes children to hold fast to it, and sanctions those who do not. This or that way, ontologically freedom is no more real than ether, even if this concept might help us in social affairs, just like the assumption of the existence of ether was a construct useful for physical theories.

There is a problem with the determinist interpretation of experimental results that point out the seeming disconnect between conscious decision and physiological mechanisms that underlie acting in the world. Libet himself suggested that his findings should not be interpreted as denying the reality of free choice.[118] Conscious will, which shows up after the readiness potential has been detected, appears before the musculoskeletal system is activated, so there is a possibility of conscious veto: the readiness potential, after all, is just that – readiness to undertake specific action, not the action itself and not a decision to act in a certain way. The existence of such veto power has been indeed found in another, more recent study: after readiness potential has been detected and up to a certain point before the action is carried out, the subject can consciously make a decision to act differently from what the readiness potential suggests.[119]

Yet this is not the only issue with the deterministic interpretation of Libet's results. Looking at readiness potential, found to be a predictor of simple action like moving one's finger, and concluding from this correlation that there is no decision is ignoring most of the possible factors that can contribute to action and focusing on something that has been chosen with a specific theoretical framework in mind. It also ignores the interaction of multiple factors, the whole that emerges when the parts are combined. This is the main flaw of the deterministic accounts that deny free will.

Determinism claims to offer a scientific explanation of what seems to us as free choice. It bases its conclusion on observable phenomena, its theory is integrated with the findings of biology, chemistry, and ultimately physics, and it makes assumptions regarding human behavior that are common to all natural sciences. Determinism cannot yet offer a theory that would explain what we usually refer to as choice. It instead assumes that there should be no relevant difference between an electrochemical reaction *in vitro*, movement of a hand, and a decision to join a marching band. It is this assumption that is the driving force of the deterministic argument. As such, philosophical determinism is based on rational argument rather than on empirical findings.[120]

Yet is this argument valid? Determinism establishes its conclusion by looking at the basic components of material objects. From observing molecules, atoms, and, in some cases, subatomic particles which are subject to quantum effects, behaving in a certain way, it concludes that larger bodies combined from them would behave in a similar way. This is logically invalid and constitutes the fallacy of composition. Each and every part of an air carrier's body is heavier than water and will sink, yet the air carrier itself would not – you cannot always conclude from the qualities of the parts to the qualities of the whole. Similarly, it is fallacious to conclude from the qualities of parts and processes that combine into an agent's body that the qualities of the body will be the same as those of its components: new qualities might emerge as an outcome of the specific composition of parts.[121] No determinist would dispute the claim that voice is different from the components of the vocal cords and the accompanying neural machinery. She would instead note that we can describe how all these components can give rise to the capacity to create and modulate air flow to issue vocal expression of different strength and pitch. This sort of explanation is needed in every case we want to conclude how the qualities of the whole are determined by the qualities of its parts. So, the determinist would conclude, it is up to those who oppose determinism to describe how undoubtedly physical nature of what we would normally consider free agents gives rise to free choice. As I argued earlier, the libertarian account fails to provide such a description.

This defense does not absolve determinism from its fallacy. However, it presents a challenge to anybody who wants to build an account of free choice: such an account should provide a coherent conceptual picture of the *how* of freedom, or describe the way whereby physical entities of a certain sort can be meaningfully free, in the sense addressed earlier under the desiderata of freedom.

No Self Account

Another challenge to free will, one that is rarely considered by the philosophers of freedom, is the argument for the illusory nature of the self. The notion of self is strongly intuitive, it is hard to construe the world or to express meaning without this notion. Metzinger, one of the leading proponents of the no-self approach, writes: "In conscious experience there is a world, there is a self, and there is a relation between both – because in an interesting sense this world appears *to* the experiencing self."[122] Such self is also necessary for any meaningful notion of freedom, as it has been argued earlier. Yet there are arguments against the existence of the self.

No-self accounts deny the existence of a subjective self, a self that is a subject that can have psycho-physical attributes and have a sense of ownership over them which is expressed through considering them as 'mine.'[123] The idea is quite ancient and has been first suggested by Buddhists: Dhammapada states that all phenomena (*dhamma* / धम्म) lack self.[124] This is because when we look at the sense of self, it inevitably relies on a variety of features and things we feel as owners of: thoughts, body, material objects, etc. Yet take these away, and nothing is left: self is constructed of them rather than owns them, it is not ontologically independent of what it mistakenly perceives as "owning."[125] Metzinger develops this further and presents several arguments against the existence of self.[126] Firstly, he notes that self cannot be established as a substance that goes beyond a body that is distinguished from any other body by the Leibnizian Principle of the Identity of Indiscernibles. If selves are mere collections of properties, then the classical Buddhist argument mentioned earlier would apply: take the properties away, and there is nothing left; hence, there is no self. Yet the most important point here is that we cannot locate the substance of the self in the brain, the seat of our neural activity: "nothing in the brain or the self-conscious biological organism as a whole could even remotely count as a substance in any philosophically interesting sense. We just don't find a substantial self anywhere in the world and nothing on the level of scientific facts determines our metaphysics in this way."[127] We have an introspective phenomenology of substantiality, no more than that. Secondly, the self is scientifically unknowable: the self cannot be accessed or observed. Only what is assumed to be self's manifestations through interaction with the world can be known, yet they prove the existence of the self no more than lightning establishes that Zeus is real. Therefore, self-consciousness cannot be seen as acquaintance with some existing entity but should be more parsimoniously considered as an illusion, together with the whole of Cartesian phenomenology.[128] Thirdly, there is a methodological ground to deny the existence of selves: "nothing in the scientific investigation of self-consciousness commits us to assume the existence of individual selves."[129] There is no empirical data that would point to self; first-person reports are no data in the rigorous sense, per Metzinger. He also notes that there is no coherent philosophical theory of first-person perspective.[130] Lastly, there is a semantic basis for anti-realism about the self: there seems to be no ontologically fundamental entity the indexical expression 'I' would refer to, there is no part of reality that would be object-like to which the expression 'I' can point.[131]

Based on all these considerations, argues Metzinger, we ought to conclude that selves do not exist. What we mistake for selves are self-models that we do not recognize as models. This phenomenal self is a process, not an entity – and thus not an agent, certainly not a free agent. By having this process an organism, a conscious information processing system, creates subjective experience of first-person perspective. Such perspective serves an evolutionary role.[132] However, it is an illusion. Just like Müeller-Lyer illusion is an unfortunate yet not too bothersome outcome of the cognitive patterns that enable us to perceive perspective, the illusion of self enables us to build complex social relations useful for our survival. Yet it still does not point to an existing self, and certainly not to a self that can be active in the world and capable of choice.

The no-self approach poses a challenge for any argument for freedom. To answer the concerns raised by the no-self account, a successful theory of freedom will have to develop a notion of the self that would be objectively accessible and knowable. As such, it cannot be based on introspection, nor can it appeal to intuitions. Intuitiveness, as Metzinger notes, is a property of claims that makes them appealing to a specific architecture of a certain system, in our case – the human mind. Therefore, intuitiveness of this or that theory cannot by itself make it correct or mistaken, just like the fact that a certain wave length falls within the human visual spectrum does not give that wave more reality.[133] Moreover, an account of the free self should be fundamental enough for the expression 'I' to point to, should provide this expression with a referent. It will also have to deal with the changing nature of the self's attributes and build a conception of self that would incorporate their transitory nature. Only such a self can later be established as a locus of freedom.

Compatibilist Freedom

Compatibilism is an attempt to establish that free will is compatible with determinism and develop a notion of freedom that will accommodate both the deterministic nature of the world and the freedom of agents.[134] These two goals can be addressed together, by building an account that would encompass both of them, or separately, namely, addressing first the possibility of freedom's coexistence with determinism, and then thinking *how* freedom can exist in the determined world.

The allure of compatibilism is clear: it allows us to accept the scientific picture of the world while retaining the view on choice that is fundamental to our culture and society and is working for us; with that we are also not falling prey to the determinist's logical errors. The commonly seen challenge of compatibilism is to explain how this would be logically possible, or non-contradictory. All hitherto suggested compatibilist accounts try to answer this challenge.

The most common way for a compatibilist account to accommodate both determinism and freedom is to re-define the latter. Hobbes, for example, defines freedom as a relation between agent's will and action: "that I can *do* as I *will*"[135] or "have a Fancy to"[136] – a relation characterized by "*absence of all the Impediments to action, that are not contained in the Nature and intrinsical Quality of the Agent.*"[137] This greatly simplifies the picture: lack of freedom becomes the constraint on individual will that prevents the agent from carrying it out, and freedom – the ability to carry out what the agent wants. The bodily aspect is clearly present: freedom dwells now between the body of the agent, objectively verifiable spatio-temporal entity, and its actions in the world in which the agent can be active. Hobbes specifically notes that freedom can be applied only to bodies, as an entity should be subject to motion in order to be subject to impediment;[138] something that cannot act cannot be restrained in its actions. This approach accommodates the consideration of causal necessity. If somebody is thirsty, it is by necessity of the way one's organism works. If one knows how to quench the thirst, this knowledge is clearly caused by the environment: the thirsty fellow had learned about water, glasses, etc., from others and from observing the environment she was placed in by a series of causes independent of her choices. Thus, the formation of the will to reach for a glass of water has clear causal necessity. Yet, on Hobbes's account, as long as she can reach freely for a glass of water, she is free; if the agent, on the other hand, is chained to her chair, being held back by other people, or is otherwise prevented from carrying out her will, she is not. Leaving the formation of the will and its nature outside the scope of interest of the discussion on freedom makes it coexist harmoniously with necessity. It also extends freedom to all entities that have a will of some sort, whether it involves reasoning characteristic of verbal humans or not.

How would Hobbes address the experience of "Children, Fools, Madmen, and Beasts," whose actions can be said to be spontaneous, i.e., originating within their organisms, but are usually not seen as free, since they do not involve deliberation, are motivated by urges or instincts?[139] Firstly, he acknowledges that the freedom to do as one wills "may be applied no less to irrational and inanimate creatures than to rational."[140] Yet even if deliberation is required, mute beasts have a chance: Hobbes notes that non-verbal animals do deliberate in a relevant sense: horses speed up to avoid painful stimulus inflicted by spurs, for example.[141] To do that, they need to distinguish between what's good and what's bad for them, and opt for the greater good. With that, there is no reason to think that these actions are less necessary than, say, drinking water in the earlier example. Same considerations can be applied to non-verbal children and mentally deficient adults. Moreover: actions that once had been deliberated upon and planned, can turn into a habit in reasoning human agents and performed without thinking, performed as if out of necessity.[142] Therefore, concludes Hobbes, the argument from deliberation is not sound: deliberation and necessity, the global necessity of all events past toward the outcomes in the future, his Concourse of Causes[143] can coexist.

It is interesting that Hobbes allows for the "Dictate of Judgment, concerning the Good and Bad that may follow on any action"[144] to be part of the cause and have some influence on the effect, e.g., as when the last straw breaks a camel's back – yet he does not explain how it would work. Perhaps, a hint can be given in his discussion on consultations.[145] Hobbes argues that from the possibility that a certain event was necessitated it does not follow that discussing it and arriving at a certain conclusion is not part of the chain of events that leads to it. If we assume predetermination of all events, my decision to jump from a tall building would be the cause of my death in the causal chain of events, and my decision to refrain from acting this way would be to. My station in this chain of events, my deliberation whether to drink the amount of alcohol requisite for the first option might or might not be necessitated – what is necessary for my freedom is the absence of obstacles on whichever course I pursue.

The problem with Hobbes's account is that the freedom it offers does not seem to be meaningful. He leaves the formation of the will out of the equation by stating that "to say, I can *will* as I *will*, I would take to be an absurd speech."[146] However, as his critics ask, if one is under hypnosis, how can one be free?[147] Perhaps Hobbes can argue that this is similar to chaining one to a chair, yet on a closer examination the case is different: the will is still the agent's will, originating within the agent. Furthermore, as Hobbes maintains that being '*free from Necessitation* [. . .] no man can be,"[148] one can argue that hypnosis is not principally different from necessitation by, say, socialization or prior actions like intoxication or obtaining certain information that would necessarily lead to particular action in specific circumstances. If only absence from constraints is considered, then a person falling from a bridge is free to keep falling, even though there is no alternative for him whatsoever.

Hobbes's account of freedom ignores choice. This makes it compatible with causality, of which he conceives as necessity, yet prevents it from developing a meaningful notion of freedom. His free agents cannot do as they please because they have no choice in the matter: the moment of choosing logically precedes the formation of the will to action.[149] When such will is formed, it is indeed the matter of constraints, yet here it is not principally different from a person in a free fall – unless, of course, this person can have a choice in the matter of the course of his descent.

Another way to reconcile freedom and determinism by re-defining freedom is to reduce freedom to knowledge. If we explore Spinoza's *Ethics* beyond the usual interpretation that stops at Part I, we will discover that this is his approach.[150] Spinoza argues that all events are necessarily causally determined, as it has been mentioned earlier, as far as they are considered part of Nature/God – and it is impossible to consider them outside of nature. However, he also notes that humans think[151] – form ideas, or conception of things mental, including but not limited to reflections of observations, using their mental capacities.[152] The "order and connection of ideas are the same as the order and connection of things"[153] – since the "the idea of what is caused depends of the knowledge of the cause of which it is the effect."[154] From here, thinking and extension are merely two modes of the same entity: we can consider the necessary order of causes and effects in thinking or in observation.[155] However, through thinking we can understand the chain of causality adequately, have adequate ideas that reflect the laws of Nature/God,[156] or fail to understand the way things are, have inadequate ideas. Without understanding the causality of nature, including one's place in it as part of Nature, human beings are subjected to emotions and belief, they perform actions of which they are completely ignorant. This, for Spinoza, is bondage, lack of freedom.[157] Inadequate ideas are not causal: they do not grasp the causes. Adequate ideas, on the other hand, Spinoza sees as causal – they grasp the real causality, and thus reflect the order of things. In the case of inadequate ideas, beings endowed with reason do not use it properly and thus lack knowledge about the state of the world. They might be worse off than beings without reason whatsoever: reason will cause negative emotions, e.g., fear, which would be strong and very unpleasant.[158] However, when one understands the causal order of the universe and our place in it, he will not be fearful, will not be tormented by negative emotions. He will be akin to a person who knows when and from which station the train departs, and when he boards one, he does not feel like he has been thrown there by a blind force of chance but rather as somebody following the set and predictable course of events. And this is freedom, for Spinoza: knowledge that brings contentment. Spinoza's free person understands that everything follows from the laws of Nature, works to obtain knowledge of these laws, and thus is devoid of hate, anger, envy, pride and similar negative emotions.[159] This reminds somewhat of Bhagavad Gita's ideal of a person who knows his place in the universe and acts accordingly, without care for the fruits of his work.[160]

Spinoza's freedom is certainly compatible with determinism. In fact, it piggybacks on it – it is essentially the knowledge of how the deterministic world works. Yet it suffers from the problem similar to Hobbes's determinism: it lacks choice. To use a metaphor suggested earlier, it is unclear how the person who knows she is boarding the right train can choose which train to board. Or, to be more precise, it is unclear how Spinoza's account leaves any place for such choice.

The question of free will is frequently tied to the question of moral responsibility. The interest in this aspect of freedom is motivated by a strong desire to maintain the traditional scheme of assigning praise or blame for actions based on the assumption of free choice by the agent. If determinists are correct, however, the assumption becomes problematic: if stealing or sacrificing herself for saving others was not up to the agent's choice but was determined by causes that precede agent's birth, then how can we assign praise or blame?[161] I find this confusion of moral responsibility and free will discussion misguided. Firstly, our wish to preserve moral responsibility should not cloud our judgment regarding freedom, as it frequently does. Our willingness to continue reasoning about morality the way we do now cannot serve as a justification for this or that position or even focus on this or that aspect of the problem. A position chosen because it serves us well by catering to our existing prejudice and thus supporting the existing order of things can still be incorrect; examples of such cases abound. Secondly, it is possible to justify praise and blame within a deterministic framework, for example, by seeing them as reward and punishment that lead to behavioral modification.[162] However, the analysis of what motivated philosophers to develop a compatibilist account might yield some suggestions that can be analyzed as accounts of compatibilism regardless of such motivation.

Many compatibilist accounts of freedom are relevantly similar to Hobbes's and Spinoza's. Hume's approach, for example, is much like Hobbes's: his freedom (liberty) is "*a power of acting or not acting, according to the determinations of the will*"[163] – where the will is tied in the chain which we conclude to be causal based on the observations we make of events as they take place in time, similarly to how we conclude about regularities of nature. Strawson's account sees responsibility for actions as a function of "reactive attitudes"[164] like good will, affection, esteem, contempt, indifference, malevolence which accompany actions that are usually addressed in moral treatment. These attitudes are what triggers responsibility. This account is similar to Spinoza's, and it is as unclear in Strawson's case as it is in Spinoza's how choice figures out in his scheme and, therefore, how we can talk here about meaningful freedom. If, on the other hand, we consider hierarchical accounts of responsibility that address second-order desires, desires to have or not to have certain desires, we encounter Hobbes's problem, twice removed from acting: there is still no adequate description of choice.[165]

THE CHALLENGE OF COMPATIBILIST FREEDOM

The main critique levied against the compatibilist account of freedom is that it attempts to resolve the contradiction between freedom and determinism by a clever definition, trying to present wordplay as a real solution. Moreover, compatibilists can be charged with re-defining freedom in such a way that it loses essential components evident from the common use of the term, and strays rather far from any reasonable set of the desiderata of freedom. Kant famously called these approaches a "wretched subterfuge,"[166] an expression that became a rallying cry for those who reject compatibilist accounts as fraudulent. He argues that a wound-up clock moves its hands by itself, having the causality within it, so on some compatibilist accounts it is free, which is absurd. Kant's free being, on the contrary, would be a rational being who, being "conscious of himself as a thing in itself, also views his existence *insofar as it does not stand under conditions of time* and himself as determinable only through laws that he gives himself by reason."[167] Similar critique of compatibilism has been advances by James,[168] van Inwagen,[169] and others.

The challenge of compatibilism, then, becomes to explain *how* a material, and thus determined being can become an agent that is the source rather than the patient of determinism, not merely as a spatio-temporal locus of causality but as a willing, intending agent who makes decisions. In other words, how can such agent if not fully reach then at least approximate some measure of absolute freedom assumed by the popular account, approach it in a meaningful way, by sharing in its practical aspect rather than having an illusion of it. This requires understanding "*how* it might be so, and how it might have come about."[170] In other words, what is required is an account of the conceptual framework of the possibility of freedom and its evolution.

An account of the conceptual framework of freedom must address its possibility, describe how it is feasible given the observable nature of the potential free agents. It should describe the conceptual pre-requisites for freedom and link them to biological enablers, since the entities that are suspected of exercising freedom are biological entities dwelling and acting in a material world. It also should address the principal differences between different types of freedom, or provide its differential characteristic: this will help to clarify its main features. This will be attempted in the next chapter that gives a descriptive account of non-verbal animals' choice vs. human freedom.

The descriptive account of the features of freedom should be supplemented by a developmental, or evolutionary account of how it might have come about. Such account would address the development of freedom in the branches of the tree of life, as well as an ontogenetic account of freedom, its development over free individual's lifetime. This will be attempted in Part II.

Chapter 2

Differential Characterization of Freedom

This chapter explores the position of non-verbal freedom in contrast to both the lack of freedom and the discursive freedom characteristic of agents endowed with linguistic intelligence. For this purpose, I will discuss the conception of life as a necessary condition for any possible self-determined choice. Then, employing the concept of life form as relevant to the question of choice, I will look at plant, animal, and human life, and the different kinds of freedom associated with them.

THE HOW OF FREEDOM: LIVING FREEDOM

As it was established in Chapter 1, in order to explain the possibility of freedom, there is a need to explain how it is possible for certain physical entities acting in a physical world, determined entities acting in a determined world, to be sources rather than merely objects of causality – to determine their own choices. Explaining this, notes Schneider, is the core of the problem of self-determination.[1] The key to providing such explanation is life. This is because life creates a self, which is impossible for inorganic nature, and, by being subject to evolutionary dynamics, enables further development necessary for freedom, namely self-control and self-awareness.

Life as Self

The first component of freedom as self-determined choice is the self. Therefore, to understand the natures of choice, different forms of choice, we need to have a minimal conception of the self that would enable choice.

William James famously saw self, "*[i]n its widest possible sense*," as a "*sum total that* [the entity whose self it is – MY] CAN *call his.*"[2] By this, he meant not only an ensouled body which can call itself its own, but also immaterial entities like thoughts belonging to this body mentally, socially sanctioned possessions like clothes and houses, etc. At the first glance this definition seems too narrow, focused on the human self as it is culturally defined in Western societies. However, it has the main component that is essential to any self – the *mine*, the aspect of belonging which can be understood not only and even not mainly in a possessive sense but first and foremost in the sense of having a clear boundary between the self and other entities that are not part of the self. This boundary, however, cannot be externally imposed, e.g., demarcated by another self – in this case it would be arbitrary, perhaps as arbitrary as political boundaries of countries drawn by former colonial powers. The boundary needs to be inherent to the entity that is to be accorded selfhood, determined as a result of the inner workings of the entity whose selfhood it demarcates, be a result of something that belongs to this entity. Such boundary would result from the activity *in* the entity whereby it relates to itself, thus asserting its identity: it is characterized by the "origination from and within" the entity itself, not an abstract identity stipulated merely by external factors.[3] The world in which the self exists and from which it distinguishes itself is the self's world as a result of the activity of the self: the way it interacts with its environment, the interfaces it maintains are determined by the self's structure and activity – and they are what constitute the distinction between the self and its world.[4] It is this kind of self, a self-relating entity that generates the boundary between itself and the world that is necessary if we are to have self as a foundation of self-determination – if no such thing is found, self-determination would be impossible.

Life answers the call. Being alive, as Kant establishes in his *Critique of the Power of Judgment*, is being cause and effect of itself.[5] Here we have two types of causality acting together. The first is a connection between two things or events where a certain state or a change thereof in one triggers a certain state or a corresponding in such state in the other in accordance with the law of nature.[6] This type of causality is the same for non-living and living entities: gravity affects all matter, tigers and rocks alike; chemical processes in living beings are more complex but are ultimately reducible to the same chemical reactions we observe in inanimate matter. The final causality, where the aims are set up by an entity and it acts in order to achieve these aims, Aristotle's final cause,[7] is usually seen in the context of rational beings who set goals for themselves to achieve, yet, as Kant demonstrates, it can be detected in its simple form in any living being. Kant gives an example of growth, noting that while the materials for growth come from the outside, and the growth itself is subject to mechanical laws that apply to all physical matter, the course of growth, the way a living being separates and combines the nutrients necessary for it, and the essential characteristics of the final product are found in the "originality of the capacity for separation and formation in [the particular] sort of natural being."[8] In other words, the end result in encoded in the very specific nature of a living being which, as Heidegger suggested, is characterized by "*self*-production, *self*-regulation, and its *self*-renewal."[9]

These causalities seem quite different, yet a self-organizing, or living, being[10] combines them. This it does through three features that characterize any life: metabolism, organic unity, and reproduction.

Living beings maintain themselves as individuals by absorbing materials from their environment and turning them into the building blocks for the organism itself according to this organism's own rules.[11] These rules are organism's own because they are encoded in the genetic makeup of each specific organism, from amoeba to the most complex primates, and instantiated in the very structure of the organism: unicellular organism membrane's functioning, gorilla's teeth and digestive system, etc. Each of these rules is a chemical process the likes of which can be seen in inanimate nature as well, and all of them are combined from atomic interactions common to all matter. However, their particular instantiation belongs to each organism, similarly to how a particular combination of paint belongs to a specific painting, even though the paints themselves are used for a variety of applications. So far, this is similar to the way non-living entities are formed: the specific processes that cause a mountain to rise or a river to form are, after all, instantiations of general physical and chemical processes in a particular area of space at a particular stretch of time. Yet the chemical processes[12] involved in the continuous reconstitution (to use Kant's term) of living beings are different in their character from the physical processes of the accumulation of matter in non-living entities, e.g., rivers, as well as from chemical changes happening to the components of non-living entities, e.g., corrosion. Inorganic natural bodies come to be and are maintained by environmental factors alone. For any such entity, its own structure, e.g., a crystalline one, is certainly a factor in the way it will develop: the way of accruing additional matter is determined by how new particles can be connected to the existing atomic construction. However, this influence is exercised through general physical laws alone, without any individuating set of processes that are embedded in the crystal and not in its environment. There is no embodiment of any sort of its own developmental algorithm in the crystal, its growth owes everything to the physical laws that are instantiated in the very same way by it and its environment – a crystal is merely a focal point for the accrual of matter, a playground of efficient causality. Similarly, the artifacts do not cause themselves but are brought about by the imagination of an external agent who also creates them by imposing imagined form on matter: the source for their formation is external, and the causality is merely final.

Differently from both inorganic nature and animal-produced artifacts,[13] organismic chemical processes are initiated, maintained, and regulated by the internal program, an algorithm embodied in living things using the materials of which they consist, of which their results are components. This is the most distinct feature that positions metabolism apart from any physically or chemically similar processes, be it thermonuclear reaction of the Sun or internal combustion of an engine: the entire material content of a living body is a result of its metabolizing.[14] Materials absorbed from the environment through metabolism are used to build the very metabolizing organism including its structures that are responsible for the selection and processing of these materials, i.e., for metabolism itself. Same applies to the set of rules defining the way in which the organism selectively absorbs materials from its environment, processes them, and rebuilds itself by utilizing them – these rules are embedded within the organism itself, in its material structures that are rebuilt using the physical components supplied through metabolism. This, according to Jonas, yields an ontological concept of the individual, or the self, as opposed to the phenomenological one,[15] the concept of a living being as self-constituting, a cause and effect of itself as opposed to merely being experienced as such by a being endowed with senses and ability to classify its environment. The latter approach would have trouble distinguishing between a humanoid robot or a robot resembling a mule in its looks and functioning, and real, living humans and mules. The former, ontological concept focuses on what the entity is, and would clearly distinguish a mechanism programmed by humans to look and function in a certain way from an organism with its own internal causality. Such living organism is the cause of itself as far as it builds itself through metabolism, yet it is at the same time and by virtue of the same set of processes the effect of itself, as the one being built and re-built.[16]

Another feature of living beings is functional reciprocity between their different parts.[17] This reciprocity exists at two levels, where the second supervenes on the first. At the first level of functional reciprocity, organism's different parts, or organs, survive and flourish due to each other's functioning. The parts also generate and re-generate each other[18] by providing the necessary material inputs: the trunk of a tree delivers raw nutrients to the leaves, where they are processed by photosynthesis which supplies the building blocks for both the leaves and the trunk. In more advanced organisms, the functioning of different organs serves each other in more complex ways: muscles enable movement which, in turn, gives access to nutrition; nutrition, processed by specialized organs, provides for the functioning of muscles. This complementarity creates an organismic unity that is responsible for organism's survival. It is also an outcome of this unity that, through simple or sophisticated bodily devices, functional feedback flows from one organ to another. Even a unicellular amoeba "senses" salinity in water and behaves in accordance with its internally encoded program – moves away. Similar mechanisms enable many organisms to try and compensate for failures of particular organs by activating other organs or entire bodily systems.[19] Unusual growth patterns in plants in order to allow leaves' access to light that is prevented by physical obstacles, development of better-than-usual hearing in individuals deficient in eyesight, or stress response where activation of multiple systems and organs following failures communicated by feedback systems can serve as examples.

At the second level of functional reciprocity, the role of a body part in the functioning of an organism, its usefulness to organism's operations in its environment is determined by its relation to the organismic whole,[20] to its unity. Such dependence of the usefulness of elements on their relation to the whole is not unique to living beings: a brushstroke can be understood as more than a colored surface only in virtue of it being part of a painting. What is unique to life is the determination of the use-inducing reciprocity by the functional reciprocity at the first, physiological level. A toucan's intricate coloring results neither from an artist's imagination nor from a general natural process affecting a certain stretch of space, but from the reciprocity of toucan's organs, their role in its metabolism, and the particular way in which this metabolism is chemically realized. In other words, it is a result of the specific way in which a toucan is a cause and effect of itself.

This way, organic unity that relies on functional reciprocity of organs constitutes a nexus of efficient causes – living being's organs cause other parts of it immediately, as a necessary condition for their generation and functioning.[21] Yet these organs are also organized in one body, and only within this structure they can both function as serving each other and understood as organs of one whole.[22] Through the functional reciprocity of organs in the organic unity of the body an organism can be seen as its own cause and effect: each part of it is both the cause and the effect of other parts, and their interrelation constitutes one unity of which they all therefore are causes; at the same time, all of them owe their roles as organs of a body to their particular arrangement within this unity, thus being the effects of it.[23]

Reproduction brings about new individuals that bear the characteristics of their parents, even though with certain, mostly slight, changes, be it prokaryotes reproducing through binary fission or sexually procreating mammals. This resemblance ensures the continuation of species to which individuals belong. Thus, in reproduction living beings figure as the cause and effect of themselves as species being,[24] ensuring the continuation of their essence, here seen as a bundle of morphological, behavioral, and other characteristics.

Living beings can be seen as cause and effect of themselves not only in terms of preservation of their characteristics but also within the framework of evolutionary phylogenesis. Each living being limits the further scope of evolutionary variations by having its offspring carrying further its genes in this or that form. Jonas likens it to a game of dice where each throw imposes certain constraints on all subsequent throws;[25] Dennett's design space, a set of evolutionary paths open for an organism restricted by its current makeup, develops this approach further.[26] The development of certain bodily systems imposes constraints on the ways in which the environmental conditions are going to impact the survival chances of the organisms equipped with this system. For example, increased digestive efficiency enables consumption of more food and thus the development of homeothermy, or warm-bloodedness, which, in turn, sets an evolutionary course different from that of creatures capable of surviving without food for a long time and thus enabling ectothermy, reliance on external heat sources, and poikilothermy, the ability to function in a wide range of temperatures,[27] as a reasonable phylogenetic path. At even more basic level, certain arrangements of genes can preclude the development or activation of other genes, effectively foreclosing a variety of evolutionary paths altogether, just like getting twelve in the first dice throw of a series of three throws precludes eight, twelve, or fourteen as possible sums of the whole series. This way, through reproduction that both passes on organism's characteristics and allows for a certain degree of variation, each living being is both the cause and the effect of itself seen not only as species being, designated as such with a degree of abstraction from evolutionary development, but also as a member in a particular branch of the Tree of Life.

These features of life are interconnected. Metabolism requires organic unity: it is impossible to imagine a metabolic process that would be sustained by a uniform entity lacking functional and material differentiation into parts. Metabolism, at the most basic level, is an exchange of matter between the organism and its environment.[28] Selective absorption of matter and its processing cannot be sustained by something akin to a river turn: no processing is done by it, only simple accumulation of material. Metabolism requires differentiated systems with specialized parts responsible for different steps in the process. Organic unity, in turn, requires metabolism. Maintaining organism as a unity where all parts are functioning in harmony and where certain organs can compensate for the failure of other organs requires a coordinated set of biochemical processes that will ensure proper intake of nutrients, their processing according to the organism's needs, and proper distribution between its organs – all in a way that ensures proper maintenance of the biochemical infrastructure that underlies the reciprocity of organic functioning. The same applies to the disposal of waste: the metabolic processes that govern the exchange of materials and energy between the organism and its environment are necessary to make sure that the waste is disposed of, yet the nutrients are retained and processed as required for organism's survival and for the functioning of its parts. Reproduction requires both metabolism and organic unity. Reproductive processes are encoded in the organism and carried out by its organs. Multiple systems are involved even in the asexual reproduction of unicellular organisms, and procreation would not be possible without organic unity and the reciprocity of organs' functioning. Furthermore, as an organismic process, reproduction requires metabolism to sustain it. Reproduction also requires metabolic change, thereby the energy needs of the organism are aligned with its reproductive functioning. For unicellular organisms, it means absorbing more nutrients and energy and having biochemical processes in the cell adjust to reproducing. For more developed organisms, e.g., plants, that would require more significant changes to support producing flowers and seeds, and then having seeds demonstrating their own metabolic processes during germination.

Interestingly, it seems that neither metabolism nor organic unity necessitate reproduction, even though empirically the living beings we encounter are characterized by all three. An individual organism can have no reproductive capacity and still exhibit both organic unity and metabolism: it can have multiple reciprocally functioning systems, absorb materials and energy from its environment in order to continuously rebuild itself, and yet not reproduce. Since organisms of this kind can leave no posterity, we would not be able to observe them routinely, with the exception of individuals with impaired reproductive systems that belong to normally reproducing species, or hybrids that, due to inheriting a certain genetic makeup, cannot have offspring. However, they still have species being, inheriting all their characteristics from their biological parents. Another case would be living beings that lack species being, i.e., those that do not result from reproduction. One would be very lucky to find such organism coming into being through, for example, amino acids and then proteins being combined into a living cell in some sort of Oparin's primordial soup. The probability of such observation if extremely low, even though it is not excluded in principle.

These three characteristics of life, interrelated and mutually supporting, can be conceptualized under the heading of *autopoiesis*. Thompson, following other researches, suggests conceptualizing life as *autopoietic*, or self-producing.[29] Metabolizing, organically unified, and reproducing, living being defines its own individual and species boundary by exercising the algorithm embodied in its organic structure. This should not be confused with autonomy – autopoiesis is a sub-species of it. Autonomy can be operational, where a certain system of entities functions as self-organizing and self-controlling.[30] An insect colony can be seen as such system: it is self-organizing with specific roles assigned to each individual by virtue of instincts and their instantiation in the colony environment; it is also self-controlling in the operational sense: raising offspring, providing food, defending its boundaries, etc. One can argue that certain man-made systems, e.g., the Internet, can be autonomous in this sense, as, perhaps, ecosystems might be as well. However, such systems are not autopoietic, they are not characterized by "an organizational closure of production processes in the molecular domain."[31] The autonomy of an autopoietic system is determined by its own organic structure and functioning resulting from such structure being a unified organism, not a combination of environmental factors that make a large group of ants function together rather than disperse (something that might happen to an insect colony), a constellation of geological and climatic forces that makes for an ecosystem, or actions of different individuals that create a technological system which has no essence outside of the way people utilize it.

How does life, an autopoietic being characterized by metabolism, organic unity, and reproduction, make for a self? By the very criterion James suggested – having a boundary that makes it an entity essentially different from its surrounding, including other similar and different living beings. This boundary, however, has to be, as it was established earlier, not externally drawn by internally driven: it is a manifestation of the activity of an entity that relates to itself, originates from within itself, is maintained through its own activity and develops according to its own program. Life is the way of matter's organization that by the way of its functioning – metabolizing as a united organismic whole at an individual level and leaving similar posterity as a representative of a species – constitutes a self. Such self, as having causality born within its own confines, is a self that can be considered as a foundation of a self that determines itself.

Life as Providing a Potential for the Development of Self-Control and Self-Awareness

Life, by its nature, has a potential to evolve, notes Jonas.[32] This is due to two features of life. Firstly, because of the genetic variation unique to each organism, in more advanced form of life aided by the mixing of mother's and father's genes, each organism is unique. Secondly, with this uniqueness, it needs to survive and procreate – otherwise its genes will not be carried further and will not play out in new organisms. In a sense, each organism is a hypothesis that can be, in the most Popperian sense,[33] falsified, where falsification means not handing its genes further. It can be provisionally supported by surviving at the personal level and passing the baton to the next generation. Genetically close individuals, or species, are akin to scientific theories: while some of their assertions might be falsified, as long as the majority is still supported, the theory holds. Yet when there is a critical mass of individuals who do not survive, the species vanishes, given way to other, more fit groups of individuals – a sort of paradigm shift in life.[34]

In order to pass its genes on, an organism needs to adjust itself to a variety of environmental pressures. The number and quality of such pressures is infinite for all practical intents and purposes: the physical conditions in which life exists depend on geological and climatic factors, which, together with presence of other living individuals, plants and animals, impact the availability of food, the need for and the effectiveness of shelter, and more. Given the mechanism of phylogenetic development briefly referred to earlier, the route of developing toward greater complexity, including developing mechanisms that would bring about self-control and awareness, is as open as the road toward becoming simpler.[35]

To discuss the contrast between non-conceptual choice and conceptual freedom, it makes little sense to focus on individual organisms or even species. This is because of the tremendous variety of life on Earth at any point in time and the infinite variety of life if we consider the passage of time. Thus, I will proceed by addressing certain types of living beings. The characterization of such types, in order to be relevant to the actual state of affairs in the living world, needs to correspond to certain basic features of natural taxonomy, yet at the same time fall along the lines separating non-discursive choice and discursive freedom. The concept of the *form of life* will be useful here.

FORMS OF LIFE

The concept of a life form can be traced to Aristotle's account of the soul. In the second book of *De Anima* he suggests that soul is a form of a living matter.[36] Similarly to any form, it defines the way matter is organized, in a sense that is specific to the form. While a geometric form gives matter spatial organization, a type of soul (ψυχή) organizes living matter in terms of the way it develops, or changes; provides it with an end; and gives it an essence.[37] Life form in this sense, Aristotle's soul, determines what kind of change an ensouled matter will demonstrate, both in terms of being affected and in terms of acting in the world. All living beings would grow and decay, yet some will be also endowed with, for example, sensation and motility. Soul would be also living being's end, since "nature, like thought, always does whatever it does for the sake of something."[38] This formulation creates a temptation to dismiss it on the charge of being arbitrary, yet it is crucial for the question of freedom. As it has been discussed earlier, life has a causality that is internal to it rather than external. This starts at the most basic biological level, with molecular autopoiesis of a single-cell organism. Thus, its own end is incorporated in the living being itself – it is the bearer of its final cause. Aristotle's type of soul, a form of life, gives functional shape to this causality, defines its type. The way such *causa sui* works in a cactus that lacks sensation and motility yet still metabolizes fabulously well given its rather harsh desert conditions is different from the way it works in a sloth who is capable of using its senses to distinguish and pursue food, albeit quite slowly; both are still different from a human being who is capable, at least potentially, to refrain from certain types of nourishment on account of abstract universal principles, e.g., religious doctrine, moral principles, or new year resolutions driven by ideas of physical beauty or fitness. Yet even more importantly, Aristotle's soul is the essence of the whole living body, i.e., the actuality of its potential. A particular form of life gives the biochemistry of living matter its actuality, dictates how it is realized in the world. However, given the peculiarity of soul as the *telos* of a living being instantiated in its very living matter, this creates a complication: if a living organism is a cause of itself, how can a form of life determine its actuality? A way out seems to be seeing a form of life as a special case of form, a case that has a potential for freedom, self-determined choice, embedded in its very being a form of *life*.

In books II and III of *De Anima* Aristotle discusses three types of soul: the nutritive soul, the sensitive soul, and the rational soul. Nutritive soul is a very basic form of life that imparts on the living matter the functions of metabolism, food intake, and reproduction.[39] Yet any type of soul, including the nutritive one, will also impart Jonas's organic unity over the living being[40] – this is the very function of the form which keeps an organism together in terms of its structure and functioning, just like a form subsumes under itself a segment of non-organic matter in which it inheres, gives it a type that can potentially resonate with cognitive structures of a mind receptive to such form. Specifically to living beings, the type of soul determines how different parts of an organism work together in order to provide for its existence. In case of nutritive soul, we have metabolism that involves multiple parts of even simple unicellular organisms working in unison, as well as digestive systems of more advanced organisms.

Nutritive soul is the only one to inhere in plants. Animals have an additional one – the sensitive soul which gives sense perception and motility.[41] This is where we can, for the first time, talk about meaningful choice rather than internal causality that serves survival. Pursuit of food, where an animal gets information through the senses and then moves or does not move toward the food source, inevitably involves choosing between alternatives: the path toward the coveted source of nutrition, the speed, avoidance of risks like potential predators of same-species competitors or braving the odds, etc. While formulated here in a human language that hints at elaboration and rational decision making where different options are weighed using some risk-vs.-benefit criteria, the choice we are considering here can exist in a much simpler form. Even in human beings many, or perhaps too many choices are made spontaneously in a non-technical sense, i.e., without any elaboration, "on a whim." Yet these are still choices: when a hare eyes two grass patches, it has to select which one to tackle first. Non-rational, without elaboration, yet still choice originating in the hare's own organism. Perhaps, a very remote ancestor of a choice made by Gaius Mucius Scaevola to put his hand in the fire for the love of Rome, the only commonality between them being the very fact of choice originating in a living being; may be, a closer relative of a choice to grab a third ice cream without considering the consequences.

The third type of soul Aristotle suggests is the rational, or thinking soul.[42] This soul, with which human beings are endowed in addition to the nutritive and sensitive souls, is what enables us to consider things through conceptualization, to generalize over particulars and to assign particulars to categories, etc. functions of thought as opposed to mere sensory perception. What seems to be crucial to this soul's impact on the question of self-determined choice is the distinction of thinking from its object, the ability to consider contents brought up by the senses or generated by a thinking process *as* something, *in light* of something – a look from above, if we are to use a spatial metaphor.[43] This greatly expands the limits within which we can make choices: somebody capable of thinking can choose not only between different available mates but also between different ways of attracting a mate, including formulating a new way of doing it. The former requires only sensation and motility. The latter becomes possible when the choosing agent can consider ways of mating as choice alternatives. The rational soul opens up a whole new realm of choosing: choosing *in accordance with* criteria that can be applicable to many different choices. Thinking beings can formulate abstract criteria and weigh the merits of different alternatives in light of these criteria. Then the criteria themselves become objects of consideration and choice: this is what philosophy does, for example. Then we can consider the ability to choose and think of its merits – and this way, recursively, *ad infinitum.*

Aristotle's souls, while providing the basis for discussing the types of self-determined choice by living beings, leave room for development necessary to accommodate an important issue related to freedom. If an organism is capable of self-determined choice, can it change itself? Can it change its own life form? If yes – in what way? How deep would it go, e.g., what would be impact of change effected by a rational life form on the nutritive soul? Further development of the form of life concept is helpful in addressing these questions.

Wittgenstein in the *Philosophical Investigations* started to utilize the expression *form of life* (*Lebensform*) in the context of use. Describing a language, he writes: "to imagine a language is to imagine a form of life,"[44] where the language is addressed as a type of communicating intent: a set of commands and later of other types of language games, activities of living beings, ways of acting or thinking. Later, he uses form of life to address agreement about what other philosophers, including, perhaps, Wittgenstein of the *Tractatus*, would refer to as belonging to the realm of universals – agreement about truth and falsity.[45] In this realm of human activity we are also talking about way of behaving, specifically about assigning certain verbal labels. The degree of consistency of this behavior would mean an agreement in form of life. Bears and humans cannot agree regarding truth and falsity since bears, due to belonging to a different life form, do not consider such abstractions. Same would apply to infants or adults whose cognitive abilities are severely impaired. It might be possible, if we are laboring within the late Wittgensteinian framework, to imagine a group of fully developed adult people who never use truth-falsity distinction – for example, those who play the language game of giving and following orders.[46] This is where Wittgenstein's understanding of life form starts diverging from Aristotle's souls – it is hard to conceive a rational soul that would not have true-false distinction built-in. Yet Wittgenstein focuses on the use. With a certain caveat, though: the limits of what sort of move can be made in a language game, the use of one's self by one's self, resurface when Wittgenstein briefly considers hope. A dog can believe, suggests Wittgenstein, that its master is at the door – with "believes" having, perhaps, an emotional character and directness rather than reflexivity that desire for food or joy might mean in a non-verbal being. Yet can a dog hope that the master will come the day after tomorrow?[47] It is hard to imagine how this would be possible, without having the concept of a quantifiable future, of its unit – a day, and of tomorrow specifically. As Wittgenstein puts it, "the manifestations of hope are modifications of this complicated [human, discursive, rational? – MY] form of life."[48] Here we can see that with the dynamism of use Aristotle's somewhat more static conception creeps back.

In order to keep the aspect of use yet reconcile it with the limitations that are evident when we consider the way of acting of humans and animals, for example, Giorgio Agamben's distinction between *zoe* and *bios* proves helpful.[49] Agamben notes that ancient Greeks distinguished between *la nuda vita*, the fact of life common to all living beings, perhaps the biochemical features of live matter, *zoe* (ζωή), and life as a manner of existence characteristic or proper for an individual or a group, *bios* (βίος). The *bios* here might be seen as a form of *zoe*, *eidos zoes* (εἶδος ζωῆς).[50] However, it is first and foremost the use of *zoe* – by *zoe*, the organism it constitutes, itself. The two, the organism (as opposed to simply living matter, physical substance with certain chemical characteristics, e.g., a collection of human cells in a lab's freezer) and its use, are inseparable. Agamben calls upon Wittgenstein to clarify this concept, as it is similar to the way rules and games are bound together.[51] A pawn in chess is not subject to certain rules – the rules constitute the pawn, without them it is but an oddly shaped piece of wood. Yet the rules themselves are nothing else but how pawn moves. In the case of chess, one might argue that the rules are a product of human minds engaged in playing chess. The case of life is essentially different, though: life is a cause of itself, here the use the form of which is *bios* is of *zoe* by *zoe*. This way, the subject constitutes itself by behaving in a certain way. Trees build themselves quite literally by metabolizing, while animals do that by eating and digesting. Birds constitute themselves as nest-builders by building nests, acquiring certain skills, here seen as merely remembered repeatable patterns of behavior, while doing so, and propagate their species being by laying eggs. Humans constitute themselves qua humans by using language and acting in certain ways. In all those cases and at all levels, from metabolism to making decisions regarding philosophical arguments, "*[t]he self is nothing other than use-of-oneself.*"[52] The use here is a form of life, what makes living subject what it is, its essence, not just one of its faculties or exercise thereof.[53]

This approach, besides giving the form of life its dynamic nature, leads to several important consequences. Having the use of a living self by the very same living self its very essence, we establish any living being as carrying a germ of freedom – the determination of the self by the self. The way it is used is what determines the type of freedom. Thus, in life forms where the only use of the self is metabolism, we cannot talk about any choice: there are simply no alternatives to access and no mechanism to choose between them. With organisms where *zoe* is more sophisticated, has access to different options by means of the senses and can realize this or that choice using its locomotion, we can start talking about choice. In a living being that can consider alternatives, compare and contrast them, create new choices, the scope of choosing widens and the choosing itself acquires a new character. Yet in any case freedom, the ability to choose, becomes inseparable from life, first as a potential, then as reality. In this sense, life and freedom are synonymous, and forms of life are forms of freedom.

Another consequence of Agamben's view, the one he notes, is that form of life, the way *zoe* uses itself, pervades the whole being, giving new role to mechanisms that in this or that shape exist in other forms of life. This way, sensitivity to light that exists already in plants, in animal form can be used to pass signals to a centralized neural processing mechanism, thus becoming a basis for visual sense. Vocal abilities of non-verbal animals, used for expressing emotions and through operational conditioning[54] or by instinct recognized by other group members as a signal of danger, in humans become a vehicle for communicating linguistic information that has a different character. This is most prominent in living being endowed with discursive intelligence: by enabling discursive reflexivity, considering its own self *in abstracto*, as an object of thought, it has a potential to change pretty much everything in the way human beings function; this includes the possibility of deciding to stop its own existence and carrying out this decision.

A third important takeaway from Agamben's analysis is that life evolves not only at the phylogenetic level, as Jonas discussed, but at the individual level too. Since each organism is separate by virtue of having its own internal causality, the use it makes of itself is necessarily different from the use another organism, another instantiation of *zoe*, makes of itself. The simpler is the organism in terms of its material substratum and the use of it, the smaller would be the difference between individual organisms. Yet it will still be present, thus giving rise to differences that will either propagate to future generations, if conducive to their bearers' survival. This is how new forms of life come into existence: patterns of self-causation carried on and being changed over time. With organisms capable of making decisions, the change accelerates, as the spectrum of different uses becomes wider: mobile animals can migrate, choices of food give rise to behavioral differences that can impact the chances of passing on certain genes rather than others, etc. At the highest yet rung of the evolutionary ladder, choice powered by reflexive thinking can even re-shape the evolutionary process itself.[55]

The infinite variety of life originates in this flexibility of living forms, their givenness to evolution and their constant change at the individual level. Life is inherently polymorphic: there is no limit on the number of ways in which metabolism, organic unity, and reproduction can be implemented in living organisms. This polymorphism goes all the way down to particular mechanisms that support life, as these can vary both in organic shape and in tactical goals, as long as they support the strategic functions of life as described above. Metabolism can be geared toward growth, as in plants, or focus on maintaining organismic functioning, as it is the case in adult animals. Locomotion in animals can be supported by wings or legs or fins or flexible body. Sensation can rely on multiple senses, ranging from tactile to echolocation. The manner of choice, the scope of available choices, and the types of choice for the vast majority of the animal kingdom are limited by these mechanisms. The more sophisticated the mechanisms are, the farther they seem to carry their owners in terms of the scope of choice: keener senses allow for more alternatives to choose from, more flexible extremities generally enable better locomotion, and more developed reproductive systems lead to better mixing of genes and therefore overall healthier and more variegated offspring. However, any of these mechanisms always limits the range of choices as well: if an animal can slither, it usually cannot jump, as it requires radically different organs; lizards cannot run as fast as tigers, yet the latter cannot afford prolonged periods without protein-rich nutrition precisely because their metabolism enables speeding up, and this requires more energy. Yet there is a mechanism, a feature of life that, once it evolves, pushes the horizon of freedom farther than any other – discursive intelligence.

In the remainder of this chapter I will analyze a number of forms of life in light of the kind of freedom they are capable of exercising. Specifically, the analysis, following Hegel,[56] will focus on plants, non-verbal animals, and discursive humans. This discrete approach is necessary to contrast choice in animals with the absence thereof in plants, despite the latter having a distinct self, and two main types of self-determined choice between alternatives: animal choice and human freedom. In Part II a continuous approach to the development of freedom will be discussed – the evolution of freedom.

PLANTS: FREEDOM IN POSSE

Plants are an example of a common type of life that bear all the characteristics of life that give rise to a self.[57] Plants metabolize by absorbing materials from their environment, break them down through chemical processes that are driven by their own organic structure and content, and build themselves, i.e., the very structures that carry on plant metabolism. Different parts of plants work reciprocally in order for the plant to metabolize and thus stay alive and grow. This feature is instantiated differently in different plants and fungi, yet is characterized by a number of common features. The nutrients are always absorbed from the immediate environment of the plant – plants are immobile and thus do not pursue food. Plants are sensitive to their environment: to light, moisture, etc.[58] Flowers open and close following changes in the intensity of light; leaves are shed and re-grown in response to changes in temperature; mushrooms pop up after rain; and more. This sensitivity is also discriminating, as plants absorb certain compounds that can later be broken down into nutrients a particular kind of plant uses, while ignoring others. Plants demonstrate organic unity: parts of the plant that perform different roles work in unison, serving each other reciprocally. The roots of the tree absorb moisture from the soil and the trunk carries it to the leaves where photosynthesis uses water and carbon dioxide absorbed from the air to synthesize sugars – which, in turn, are used to build the roots, the trunk, and the leaves. All these parts of the plant grow, albeit in a fashion very different from that of animal parts. Plants do not have a specific shape: while the texture of wood and the shape of leaves are determined by the tree's DNA, the number of branches and of leaves, as well as the structure of branches, is indeterminate.[59] There is, in a sense, more independence to each part of the plant, there is no central control over its living being. However, they still work in unison.

Recent research produced some remarkable findings that made many question the status of plants as being lower than animals on the developmental scale drawn in accordance with the criteria of complexity and ability. For example, it has been observed that when giraffes start munching on acacia trees, the latter release certain chemicals that offend giraffe's taste buds.[60] The chemical compounds are sensed by other acacia trees in the surrounding, and they too unleash similar chemical defenses. Giraffes, capable of operational conditioning, avoid the acacia trees in the proximity of their first victim and walk farther, to find acacias that have not picked up the scent. Additional research shows that there is a forest behind the trees, or what has been called a wood wide web: trees are interconnected and exchange nutrients, either through their connected roots or with the help of mycorrhizal fungi, thus affecting a sort of resource sharing.[61] These and other similar findings have been summarized and given a far-reaching interpretation in a bestseller that arose wide interest among the public and caused significant irritation in the scientific community, Wohlleben's *The Hidden Life of Trees*.[62] The author argues that the best interpretation for these advanced abilities of higher plants is to consider them as communicating, sharing resources, etc. – imputing to plants behaviors that are usually reserved for animals, if not for humans. This trend in botany gained the name of plant neurobiology and suggests interpreting certain physiological processes in plants, in particular those related to their response to environmental stimuli, as similar in function to that of animal nervous system.[63]

The problem with such approach is that, as many plant biologists note, "there is no evidence for structures such as neurons, synapses or a brain in plants."[64] Indeed, plants lack any component that would exercise centralized control over the plant organism, integrate the biochemical responses to environmental stimuli – an integration that would allow to call receptivity sense perception, and summon the organism to a unified response. Without such structures and functioning, a leaf triggering a release of a certain chemical when chewed, and another leaf of another acacia tree that picks up the scent and responds with releasing similar chemicals, cannot be seen as organs of a responding organism. They are merely parts of a structure where they have reciprocal biochemical relation. This structure would be more advanced than a river bank that "rebalances itself" by collapsing (response) after enough soil has been washed away by the river (stimulus), but only by virtue of having common metabolism, organic unity, and reproduction – not because of having any self-control.

Nevertheless, plants are characterized by autopoietic life, internal rather than external locus of functioning as their metabolism, organic unity, and reproduction are driven by the algorithm embedded in the plant's structured matter. Thus, plants demonstrate a self. Yet what kind of self the plant's self is, what sort of life characterizes it, what type of use it makes of itself?

Hegel characterizes the plant self as possessing *formal subjectivity.*[65] A plant is a communion of individual members that work reciprocally rather than a unified whole. The identity of plant's members does not depend entirely on them being parts of the plant – differently from animal's organs. Branches of trees and bodies of cacti can be grafted on other plants; for many plant species branches can be planted in the ground and grow into new plants. Centralized control over the shape of the plant is lacking as well: the number of plant parts, as it has been mentioned earlier, is not determined by the plant itself but by external conditions to which it is subjected by virtue of its seed being planted in a certain place.

Lacking centralized control, plants are unable to act as a unity on their own determination. Not having the necessary organic control structures that would functionally resemble a nervous system of animals, plants cannot determine their location in space – it is controlled by gravity[66] which thus determines plant's access to nutrition; its growth is controlled by light,[67] and its germination – by temperature, humidity, etc., external factors. Hegel analyzes specific aspects of plant's being in three aspects of its life: formation, assimilation, and genus. The formation of the plant organism is characterized by externality: its development is outwards-oriented, it is differentiated into members that expand outwards and have incipient independence. Assimilation, or metabolism, in plants is characterized by immediate absorption of nutrients located contiguously with plant's body[68] – without such contiguity plants, being unable to move, would not have access to nutrition. Plant's genus process, or reproduction, is carried out by proliferation, by its parts getting out of themselves rather than by an organism acting purposively as a unity:[69] plants lack any means to pursue potential mates, the context of mating is irrelevant to plants. These three processes in plants are indistinguishable. Plant's formation, differentiation into members, is exactly how the plant sustains itself: it survives by expanding, which is how it metabolizes, or turns the nutrients it absorbs into its own matter.[70] Yet this is the way it reproduces as well: proliferating into new branches, buds, etc. Thus, the formation, metabolism, and reproduction in plants are merely formal,[71] the distinctions between them are more conceptual than concrete: in reality they are realized as one and the same process.

Consequently, in the case of the plant life form we cannot talk about self-determination. There is a determination that originates in plant's self, activities that are driven by its living algorithm, as in all the examples of metabolism that have been brought up earlier, yet what is lacking here is the *use* of self by the self, the determination of *zoe* by *zoe*. Instead, the build-up of the plant seems to be more of a by-product of the functioning of its parts. In self-determination, what is required is a self-focused action, where by doing something, the subject does not just build itself but constitutes itself *as doing something*. Agamben, suggesting this, cites Spinoza's comments on a Hebrew verb form *hitpael* (לעפתה, *reciprece*),[72] the role of which in English is usually expressed by prefixing verbs with *self-*, as in self-destruction and self-referencing. Spinoza notes that this verb form expresses an action where the subject and the object are one, the focus of the action is the acting self.[73] In the case of plants, since their subjectivity is formal, their being united in reciprocity of parts that depend on each other yet are not centrally coordinated, there is no action of the self *qua* self but only as self *qua* collection of dependent parts. Such action can have no focus. Plant's metabolism builds the plant as a physical entity, yet it is not exercised with the focus on the plant, purposively – there is no central control that would allow for purposeful action of the organism. This, in contrast with actions characteristic of animals, e.g., flight from danger the focus of which is the organism itself, a purposive action where all organs work in unison, controlled by the central nervous system of the organism focused on its survival.

As incapable of self-determination, plants cannot exercise choice. Their sensitivity does not amount to producing sensory data that is integrated in a way that can give rise to a purpose of action, or choice alternative. Their functioning, a reciprocal operation of otherwise independent parts, is not unified enough to make choices. Therefore, in the case of plants we cannot talk about freedom but as *in posse*. The foundation of all freedom, the distinct self, is present in plants.

As live beings, plants evolve. It is this evolution that can gradually lead to re-purposing and further development of certain features into ones enabling choice, e.g., evolving parts that are sensitive to stimuli to organs of sense perception. Some examples from biology, e.g., sponges that are considered simple animals, illustrate this transition: sponges travel very slowly, using plant-like structures as they lack a defined locomotive apparatus, have undefined body shape, are not known to pursue food and consume whatever happen to flow through them, yet seem to coordinate their functioning as they possess distinct circulatory, respiratory, digestive, and excretory systems. Thus, plants can be seen as standing on the doorstep of freedom yet not quite crossing the threshold.

NON-VERBAL ANIMALS: NON-CONCEPTUAL CHOICE

The Main Characteristics of Animal Organism

Aristotle sees animal soul having *sensation* as its principle: the ability to perceive stimuli rather than merely respond to changes in the environment.[74]

Perception, if we continue Aristotle's line of thought, would be supplying information to the organism that is capable to act upon it, e.g., based on a fixed habit or instinct. Aristotle argued that the sense organs that support sensation are absent in plants.[75] Indeed, we are yet to discover anything analogous to a human eye or ear in shrubs and pines. However, the research of plants' sensitivity mentioned above suggests that, even though the biological mechanisms of sensitivity are different, there are certain commonalities among animal and plant sensitivity. To use previous example, acacia tree's ability to pick up a chemical scent from another tree might be seen analogous to an ant or a mammal smelling a pheromone. The difference, though, is in the level of integration. In plants, different parts have a high degree of independence, as they are not centrally integrated. In animals, as Hegel notes, these parts become organs that are integrated so tightly that they are useless outside of animal organism.[76] This is what makes sensitive parts into sense organs: their outputs are integrated and then used as data for making choices. Not merely a tropic response by one part of a plant to a chemical compound picked up by a leaf but a coordinated response of the whole animal organism, e.g., fight or flight. The animal life form utilizes some evolutionary patents implemented in a more primitive form in plants, e.g., light sensitivity, in new ways: by integrating them into one system with unified control, they acquire new roles, they are now *used* in a different way, like an image that starts playing a role of a letter in a hieroglyphic writing system.

In order to emphasize this core difference between plant parts and animal organs, it may be useful to look at rudimentary remnants of the plant world in animals, unorganized as branches of a tree. Hair might provide such example: the number of hairs is undetermined. In many animals, e.g., humans, the central nervous system exercises but very basic control over them. Hairs from one part of the body can be grafted onto a balding scalp successfully. If we compare it with a finger, we would see the difference: fully integrated with the rest of the nervous system in terms of receiving and passing neural signals, with the circulatory system in terms of blood flow, with the musculoskeletal system in terms of its movement mechanisms, etc., a finger's specific role fully depends on its placement in the grand scheme of the body.

The exercise of a unified control over the organism is achieved by the central nervous system with a unique organ for which there are no analogues in the plant world – the brain. This is the place to note that any attempt to draw similarities among all animals encounters difficulties when the discussion progresses into details: it is hard to avoid principal differences in pretty much every aspect between a starfish and a chimpanzee. However, the existence of primitive plant-like forms, or even organisms whose belonging to the animal kingdom is debatable, is only natural if we are to think of the development of life in evolutionary terms. Nevertheless, the only type of living beings that are categorized as animals today that do not have a nervous system are sponges. The only ones without the brain are cnidarians, e.g., jellyfish, and echinoderms like sea stars and sea cucumbers. The rest have a centralized nervous system with a special organ dedicated to integrating inputs from sense organs and controlling the whole of the organism.

Sense perception provides the basis for another animal characteristic: desire, which in turn leads to the feeling of pleasure that results from the satisfaction of desire, and the feeling of pain that is related to the lack thereof.[77] Integrating the data from the senses, animal organism can figure out satisfaction or lack and pass control signals to change its behavior accordingly: the sensation of hunger or tiredness would affect the commands to pursue nutrition. This seems to be a necessary ability to pursue foodstuffs rather than absorbing them from organism's immediate environs. This leads us to the recognition that animal soul also gives the organism a sort of unity importantly different from that of a plant: unity that controls the organism as a whole, allows the organism to determine its actions.

The primary example of unified action in animals is movement, for which desire gives the impetus and which is directed by the results of processing the information provided by the sense organs.[78] Having a brain exercising centralized control over the organism enables motility – a coordinated movement of the whole organism as opposed to its individual parts.[79] The determination of location in animals is made by factors internal to the organism that operates within the confines of physical laws. While both birds and reptiles are subject to the same laws of gravity as palm trees and moss, the former can move following their own internal determinations, influenced but not driven by extra-organismic factors. As it is the case with sense perception, motility requires special capacities lacking in plants. Some of these capacities are further development of those already present in plants, as it is the case with sensitivity pointed out above, e.g., systems that extract energy from the nutrients are more sophisticated and efficient in animals as more energy is required for movement. Others are entirely new, as the central system and, in many animals, special body parts that facilitate locomotion: musculoskeletal system, extremities like wings, fins, and legs, and sympathetic nervous system that enables more efficient fight or flight response. As it is the case with perception, motility also increases the evolutionary usefulness of the components of animal's motor apparatus, which in turn enhance the need for a more effective central nervous system. Same relationship exists between motility and perception: more sophisticated senses, most notably vision and hearing, provide new reasons for motility, e.g., acquiring food that has been spotted at a distance or avoiding danger detected by the senses. The ability to move toward or away from a stimulus, in turn, make vision and hearing more valuable.

Bridging the gap between the organism and its food, as well as avoiding danger, requires not only the means to alter location, provided by motility, and ability to pinpoint the focus of pursuit or flight, made available by sentience. A sustained intent that will persist over time is needed too. Such an intent is emotion.[80] It enables the animal to keep focused on its goal: greed focuses on pursuit, fear – on avoidance, etc. While perception provides animals with a spatial horizon, emotion provides them with a temporal one. Motility allows bridging the spatial gap, yet this needs action sustained over time. Emotion, an urge that is informed by perceptual data and at the same time directs perception, makes it happen. It is the emotion of greed, in Jonas's terminology, that provokes the motion when something resembling a suitable nutrient is spotted, smelled, and/or heard; it is the emotion of fear that causes motion when something threatening is sensed. Yet a hungry animal would also be looking for food, directing its sense organs and giving preference to certain kinds of perceptual data over other kinds, e.g., that about the prey over that regarding potential mates.

Jonas argues that perception, motility, and emotion are all manifestations of the distinguishing principle of animal life: *mediacy*.[81] While the plants are absorbing nutrition from the contiguous environment, animals need to reach over to their food. This basic difference in the mode of metabolism gives rise to all distinctly animal features.[82] The spatial gap is also temporal: reaching over spatial gap of any significance requires sustaining movement over time. This has evolutionary significance: to survive, animals need to hone their senses and increase their ability to distinguish between different elements of their environment. For the same reason, their urges need to be strong enough to sustain pursuit and flight.

This gap is what gives birth to the animal type of subjectivity. In animals, due to the spatio-temporal gap between the desire and its satisfaction, we have the divide between the organism and its world. Plants live in a world, yet their only separation from it is because of their degree of internal determination which makes for a *formal* subjectivity. Animals *have* a world, their world exists *for* the animal organism which needs to perceive it and act in it in order to survive. In Agamben's terms, the animal form of life, its *bios*, consists in its *zoe* using itself against this world, as a subject confronting a *Gegenstand*, an object that stands against it. Perception bridges over this divide between the animal subject and its world receptively; motility – actively; and emotion sustains the movement, its correspondence to perceptual inputs, and the handling of perceptual inputs in accord with the goals it sets.

Based on this, Hegel defines the main principle of animal life as *concrete subjectivity*:[83] an organism that exists *for itself*.[84] Animals' concrete subjectivity finds its expression in a number of aspects which Hegel analyzes alongside three scales: shape/formation, assimilation/metabolism, and genus/

reproduction. Animal shape, as it was already noted, is centrally controlled, thus pervaded by organismic subjectivity: no organ can exist without other organs, and the organism itself is fully expressed in its organs. Physical structure of animals both expresses and supports this unity: animals have viscera, differently from plants that grow out of themselves and do not have internal organs; animal skeletons protect the viscera, be it the exoskeleton of insects or a mammalian osseous system. As a result, sentience can develop, realized in the nervous system that supports it, further divided by Hegel into internal and external sensation. Sentience is different from plant's sensitivity to environmental stimuli, as it has both inward and outward reference, thus supporting a rudimentary sense of an I, recognizing even though not conceptualizing feelings as belonging to the subject – subjectivity *for itself* rather than merely *in itself*, as it is the case with plants. Animal's organismic unity, ensouled by sentience, is felt as such by the animal itself, making the possession of the *feeling psyche* its main feature.[85] This psychic nature, relationship to itself, is a new element crucially important in the context of choice: it is at the core of the determinations made by the animal organism, starting from fight-or-flight decision and all the way to sexual relations and care for the offspring. Animal life then is *life aware of itself*, even though, being bereft of discursive rationality, not aware of itself *qua* life or its own self.

The mediacy noted by Jonas, coupled with self-awareness, leads to another important development: it builds the animal's I as distinct from its environment, of which the animal becomes aware through internal sensation. Hunger is *his*, felt, even though not conceptualized, as such; the banana is not, felt, even though not conceptualized, as such. The practical relationship that drives the acquisition of food in the conditions of non-contiguity of nutrients with the metabolizing entity is that of lack: feeling a lack and the urge to satisfy it.[86] The feeling is enabled by the organismic composition: only what feels, and feels itself as itself, can feel a lack.

Neurophysiologically, in many animals the process of alleviating lack is supported by instincts,[87] inherited behavioral patterns responding to the feeling of lack that specify a certain range of targets[88] yet allow for a degree of freedom in choosing the way to pursue them. Instincts are not mechanistic stimulus-response algorithms but rather loose connections that provide spectrums of response patterns to types of stimuli, complemented with some specific action sequences that can be performed without learning. Even in animal species where instinctual behaviors are seemingly precise, e.g., nest building or mating songs in birds, the inherited outline is informed by the senses, animal's life history, and its material and social surrounding, thus being expressed differently in different individuals. This is an important point in understanding instincts: they not only limit choice by dictating its parameters but *require choosing*. A bird feeling an instinctual urge to build a nest has to choose the materials to work with, to select this or that tree, this or that twig. Same applies to mating, grazing, hunting, etc. A mountain goat is hungry; its diet is determined instinctually – yet it is up to her to choose both the patch of grass to graze on and a path to reach it. Animal's life form focuses on exercising these choices; its nervous system, its motility, its senses are geared toward making them.

The genus process, or reproduction, sheds light on another aspect of animal's relation with the world. While assimilation/metabolism leads to the relationship of lack/satisfaction, in sexual relation an animal recognizes its own element in the other, its potential mate.[89] Here we can see the first step of the animal going beyond self-feeling, recognizing, although non-discursively, an abstract element that is common to itself and another living object – recognizing a life *similar* to its own in the world. While governed by instinct, sex-relation in animals can be seen as a pattern that in the next evolutionary stage will be re-purposed: it is this concrete relation that unifies individuals based on which reason would later develop abstract principles, going from the feeling of "he is like me" and "X is like Y," non-discursive recognition of commonality and thus generality, to universality.

Mind

Aristotle addresses mind (νοῦς) as a part or the aspect "with which the soul knows," the part where thinking takes place.[90] This leads to several important consequences.

Mind in its minimal form, Hegel's *psyche* (Soul/*Seele*), is merely reflective of what is brought to it by the senses and the inner feelings, e.g., desires and passions – "only the *sleep* of mind – the passive *νοῦς* of Aristotle."[91] As such, it only feels what is immediately given.[92] It also demonstrates some, however minimal, degree of independence from the externally given – a nature of its own. This is because it supervenes on the organism's self, with its differentiation from the surrounding word, selective intake of stimuli, and their processing by the organism's own nervous system, however primitive at the earlier stage of animal evolution it might be.[93] To be both independent and reflective, the psyche needs to be capable of addressing different givens and be universal in regard to the contents it can process. If it is defined by some specific contents, it can neither be reflective of the different givens served to it by the senses nor can it retain a degree of independence from them. Without being reflective of the world, mind cannot be of any help to the living organism functioning in the world: if the goat's mind cannot reflect and distinguish the reality of grass vs. rocks, this goat will not survive for long. Yet if its mind is geared toward specific inputs, similarly to a mechanism that is programmed to register and process only a set of pre-defined signals, a goat's survival would be as problematic in a world where other selves operate and environmental conditions change. Moreover, the soul will not awaken from its passive slumbers to the distinction between subject and object if not for the germ of independence, necessary for the subject, and reflectiveness, necessary for objectivity.

One can object to this reasoning by noting that the senses operate only in particular modalities and register specific ranges of physical stimuli. Different animals might lack certain sensory modalities altogether or have them developed to different degrees: roaches are deaf, bats can echolocate while cats cannot, and dogs can smell much better than humans do yet lack in vision, even though they are superior in this regard to moles. This type of limitation, however, does not impinge on the philosophically interesting universality. The psyche is universal in regard to sensory givens since it will reflect all inputs within the sensed range: it has no part in limiting the sensations, this is the province of the sensory organs and the corresponding neural structures.

This essential universality of the psyche provides the starting point for the development of mind all the way to its truly universal form – thinking. Thinking is not limited by sensory or any other contents. This is because its vehicle is language, a system of "infinitely generative grammar and indefinitely flexible lexicon."[94] As such, thinking can relate not only to sensory givens but also to itself: to contents generated by thinking itself like irrational numbers, normative statements, or philosophical arguments about the features of thinking. This stage, however, cannot be reached by non-verbal animals as they lack language; it will be addressed later when human rational freedom is discussed, and specifically in the section *Rational intelligence*.

Remembering Kant's *Second Critique*, we might say that mind can have no content of its own, even though it can impose certain forms on thinking.[95] For the same reason mind is "not actually a real thing"[96] – it cannot be associated with any specific organ, for example. At the same time, it cannot be separated from the body: mind is part of the soul, and soul is body's form, its actuality.[97] In Agamben's terms, mind would be part of a particular kind of *zoe* using itself. Therefore, it makes sense to conceive of mind as an element of life that is internal to the organism, has a kind of existence that is inseparable from the organism that is capable of supporting such aspect as long as it is alive, is separate from each and every part of the organism when such part is considered on its own yet pervades all these parts when they are considered as organism's parts. As such, mind is inherent to the organism and is not imposed on it as a way of conceptualizing its functions by philosophical analysis. From here, mind would entail subjective experience and a measure of self-control.

What kind of organisms will have mind? Firstly, to have subjective experience, an organism cannot be equipped for merely tropic, localized responses to external stimuli, where one of their parts responds to a stimulus while others do not.[98] To have experience as one subject, an organism needs unified sensitivity, where all outputs of organs that reflect external stimuli or events internal to the body of the organism are processed together – without that, being aware of itself as a self, one self, is impossible. Secondly, they have to be unified in terms of control: in an organism where each part is on its own and only participates in the community of parts by virtue of reciprocity of metabolism, nothing that pervades all parts of the organism would be possible. Thus, plants will not have minds: they have no unified control of the organismic whole. Lastly, since mind would not be a specific thing, specific content, but potentially relate to any content, an organism to have mind would need some mechanism of generating content for the mind to work with, for example, sense organs to bring impressions. Therefore, animal form of life would seem to be the first candidate for having even the simplest kind of mind: animals have neural structures to exercise centralized control over the body and senses to bring impressions form the outside.

Pervading all parts of the organism would also mean a measure of *self*-control, where the organism *acts* on its own accord, following its own psychological states rather than merely responding to environmental changes. For an organism possessing a mind, the latter are inputs processed by it, yet they do not lead to an automatic response that can be predicted with certainty based on the understanding of organism's biochemical composition. Organism's psychological states might have different instantiations, yet at the most basic level they can be what is usually referred to as *emotions*, states of the self that pertain to the organism as a whole separate from its environment, that do not require conceptual understanding: fear, excitement, anxiety, etc.

Mind is the way by which an organism endowed with it relates to its own self. It is through mind that *zoe* feels its own feelings; when it confronts objects, through mind it confronts their representation provided by its own senses; when it thinks, through mind it thinks its own thoughts.[99] As an element that pervades the whole of the organism, mind enables *zoe*'s *awareness* – centralized processing of environmental data provided by organism's different sense organs, processing that will treat it as something pertaining to the organism as a whole. In other words, mind will entail *subjective experience* – not only being an entity distinct from the environment in terms of having its own life process, but also feeling oneself as such.[100] Having a mind means experiencing the world as separated from the self and experiencing the self as a unity. This, in a being whose metabolism is conditioned upon reaching out to food – a motile animal endowed with senses – means intentionality, a minimal form of goal-directed behavior.[101] Intentionality is a direct result of organism's feeling its own feelings and being able to act upon what it senses, e.g., moving away from danger or activating an instinctual patter like salmon's urge to migrate that utilizes its geomagnetic homing mechanism.

How mind is instantiated in this or that animal would depend on the particular neural mechanisms the animal possesses. A rudimentary nervous system of a starfish can hardly claim to have a developed mind – its self-control is very basic, senses underdeveloped, and the notion of a starfish's mind would be more of a potential. On the other hand, in the case of dogs we can see a fully developed control mechanism enabled by a capable brain, advanced senses to bring inputs, and memory – a retention mechanism that can also preserve and provide contents for mind to operate upon.

Central Features of Animal Mind

The animal world is extremely diverse on the scale relevant to the question of choice. Some non-verbal animals have very limited mechanisms of self-control, primitive locomotive systems, and rudimentary senses, so it is hard to conceive of them making any choices: a starfish has so little in terms of senses and motor capabilities that its choices seem merely formal, similarly to subjectivity in plants. Other animals have well-developed self-control complete with consciousness and signaling mechanisms that approach the threshold of verbal signs: baboons have consciousness and communicate extensively, vocally and otherwise, even though without using language.[102]

There are several aspects of mind that determine animal's position along this continuum and are relevant to the question of freedom: awareness, intuiting consciousness, representing consciousness, and associative imagination.

Awareness

Awareness, as it has been alluded to earlier, is characteristic of any minimal form of mind, supported by even a rudimentary nervous system. Awareness amounts to bare feeling: registration of organism's own state, feeling its own feelings as own feelings.[103] Hunger, thirst, discomfort, etc. – an animal organism feels all these *qua* organism, a unified entity, where a signal from one organ is interpreted in the central nervous system that pervades the whole of the organism. This is what allows the animal to respond as organisms, e.g., trigger its locomotive system. This is very different from plant's tropic response, where a leaf releases a chemical in response to insect's saliva, even if such chemical causes other parts of the plant to release more chemical compounds.

Awareness in its minimal form does not require distinction between subject and object. However, it is a necessary phylogenetic and ontogenetic precondition of subject-object distinction: without awareness of own feelings it is impossible to distinguish between one's own and another's. Such basis is provided by two characteristics of a pre-conscious psyche: expressing its feelings and development of habits.[104] By having central control over its organs and feeling its own feelings, an animal is capable to give behavioral expression to what it feels, be it releasing a scent in response to fear or movement in response to hunger. This later becomes a building block for conditioned responses, as an action itself can cause a chain of events the consequences of which will be felt. By feeling repeated feelings and responding in a certain way, mind is being habituated to feeling-response pair,[105] which later requires fewer neural resources to be performed – sort of an automated response which is probably facilitated by making neural shortcuts, conceptually akin to polishing a board over which a brick has been slid many times. This too will later provide a cognitive building block for generalized sensitivity.[106]

A mind equipped with merely basic awareness will be capable of choice only formally, similarly to how plants can be considered subjective. The choices are made between alternative options here based on what the organism feels. These choices determine the organismic self: it moves, gets satisfied or not, etc. However, at this stage the organism is not yet capable to represent alternatives in order to choose one. Therefore, its automatic choices cannot be seen as meaningful freedom.

Sentient Consciousness

Animal mind's next stage is sentient consciousness. It is enabled by further development of the central nervous system which, in turn, evolves under the evolutionary pressures on animals who are capable of awareness and, due to its habituation, have mental resources to spare and re-purpose[107] – as any change in the life form, this is a change in the use of *zoe* first and foremost. Here, using the resources provided by the sensory organs and self-feeling, the organism starts distinguishing between the self and its world, or exercise subject-object distinction. An organism here registers the immediately given through the senses, cognizes it as such and apprehends that it is both a modification of its mind and an externally given object[108] – combines "the self-feeling of the psyche with the sensation of and immediate given."[109] This is an essential ability for animals whose pursuit of feed, mating, etc., behaviors involve complex patterns of activity, for example, travel. In fact, most cases of mating, migration, and hunting we can observe in the animal world are hard to imagine without the subject-object distinction.

It is important to note that the subject-object distinction in this minimal form can operate without theorizing, without language. The animal here is "aware of its object but not of the relation between its mental content and that object, whatever it may be."[110] This is because to be aware of the relation itself, mind needs the ability to treat its objects recursively: first a thing, than the relation it has to the thing, then the relation is has to this relation – potentially *ad infinitum*, with, perhaps, a physiological limit determined by the organismic central nervous system's computational power, yet without any limit imposed by the theoretical capacity of the mind. This ability comes only with discursive metacognition and language.

Still, a pre-linguistic mind can be aware of the difference between memory and perception. This can be accomplished by distinguishing between the sources of a mental representation, between what is registered by the senses and delivered for further processing as opposed to what is being pulled from memory. Such distinction is present in animal minds: it is evident from their behavior, as animals respond differently to contents they are exposed to through the senses and to contents they remember. Dogs remember the association between hissing and cat claws once they have been exposed to it – this much is evident from them retreating at hearing a cat hissing. Yet they do not retreat merely based on memory, without being exposed to a corresponding sensory experience. A failure to distinguish between remembered and perceived contents would be lethal for animals functioning in the world: a food remembered cannot sustain organismic functions any more than Kant's imagined five dollars/thalers can buy a meal, and a perception of a predator mistaken for its memory will cut short all possibility for further acquisition of memories.

While subject-object distinction is a significant step toward choice, even at this stage it is no more than a formal choice: an animal endowed with intuition focuses on the immediately given, does not retain information about intuited objects, and thus cannot make a choice between available alternatives but automatically. It is also hard to think of an animal who would be able to intuit yet not retain intuitions, even for a short period of time. However, the impressive variety of life on Earth might well provide several examples, in the past, now, or in the future.

Representing Consciousness – The Birthplace of Choice

With the addition of the ability to retain sensory images, we arrive at representing consciousness. This development seems to be enabled by a sheer increase in the neurological resources or freeing up some of them thanks to automatization of certain feeling-response pairs. Whatever its physiological correlates are, at this stage an animal, after intuiting objects, is capable of storing intuited images, for shorter or longer period of time, with more or less contextual information. Then, the animal can recollect the intuition and compare recollected images: visuals, smells, tactile sensations, or combinations of them.

Recollected images are distinct from the intuited images. While the latter are reflections of immediate presence, the former are mental products. They are retained by animal's neural system and have usefulness only as such: an animal who confuses recollected smell with one present would be hallucinating, this way severely impaired in its functioning in the world and hurting its chances to survive. The change here is a change in form, not in content: what matters is not how vivid the image is, as Hume suggested, but how does it stand in relation to the environment being intuited right now.[111] Thus, the acquisition of the capacity to retain and retrieve images means gaining a temporal dimension to animals' mind, especially if we consider association of images.

An essential element that makes recollection of images useful for animal behavior is associative imagination. The images are recollected when an image similar to the stored one is intuited – this has been known to all animal handlers and documented in Pavlov's and Skinner's research at the beginning of the twentieth century. We do not know of any organisms who are capable of retaining images yet not associate them: non-verbal animals have no other way to indicate image retention but to respond in a predictable way to a similar stimulus, where predictable response is the one associated with the initial intuition, the image of which has been retained. The philosophical significance of association is that the images are combined here in a certain order of animal mind's own construction: it is the organisms that organizes the web of associations, relying on some aspects of the intuited manifold while inevitably downplaying or ignoring others.[112]

The past dimension is clear when we think of the retention and recollection of sensory images: it is founded on the distinction between the intuited and recollected images. Yet together with it comes a future dimension: an animal, by recollecting an image triggered by a similar one being intuited now, essentially flips the past image into the future: the meaning of associating a recollected image of a tasty reward with a certain kinesthetic feeling of giving a paw upon hearing a command "Give me a paw!" is that giving a paw *will* cause a similar result. In this sense we are talking about constructing a mental representation of a future event.[113] Such construction is purely episodic, i.e., driven by recollection of images sensorily associated with the currently intuited one, yet it nevertheless points to a future state of affairs. This has a principal significance to the question of choice: any choice is based on existing information yet points toward the future.

This is the developmental point where intentionality becomes teleology: at this stage, the explanation of behavior can be given not in terms of some antecedent unobservable entity but rather as "a function of the state of the system and (in the case of animate organisms) its environment; but the relevant feature of system and environment on which behavior depends will be what the condition of both makes necessary if the end concerned is to be realized."[114] Explaining the behavior of a goat scrambling to get to a green patch on a rocky hilltop based on antecedent conditions alone, as can be done for a mechanical system like billiard balls on a table, makes little sense given that a goat is a living system that, driven by an urge to alleviate hunger, can represent a desired state of affairs – the situation where it is grazing on delicious green grass. As a motile being equipped with vision, smell, and, perhaps, a kinesthetic sense, it can also set tactical goals – getting to the next rock on the way to the coveted patch of grass. This can be done again by pure representative imagination, as it is done by non-verbal infants who are perfectly capable of advancing toward certain desired objects while overcoming obstacles.

This is the evolutionary and, in specific organisms including human children, individual developmental stage where choice becomes possible. Firstly, the animal here is capable of representing alternatives – in this case, recollected images. Secondly, a comparison may be made here between different results. The results brought up to animals' consciousness are the recollected images, and the trigger – intuited images. Thirdly, the comparison *can* be made at this stage – associative imagination has already accorded the organism the ability to link together similar images. *How* is the choice made, how it can be made without external, universal criteria? The images can be compared in light of animals' desires and their capacity to satisfy them: causing pleasure vs. displeasure, in the most general sense. Since we are considering here the comparison of images, it would translate into animal's senses and their relation to pleasure: tasty food, offending smell, frightening noise, etc. Here for the first time the animal kingdom crosses the threshold of freedom, albeit making a small step at first.

Associative Imagination

The limitations on choice vary along with the capacity of animal's mind to associate images. The more images an animal can associate, the more associative dimensions are available, the more steps animals' cognitive machinery can retain between an intuited image of the behavior of the self and the image of the resulting intuition. New experiences lead to new images stored and new relations formed. The animal accumulates more internalized content, and the repertoire of choice becomes much richer: there are more alternatives now to choose from. The gates of freedom crack open, and the gap is getting wider with every increase of cognitive capacity – capacity to retain images and to relate them. Yet expanding the horizon of choice is not limited to increasing internalized content. The expansion of sensory capabilities is also a major contributor to freedom. Both here reinforce each other: this expansion itself starts making evolutionary sense only when it can impact behavior, and this impact comes through it influencing choices. Empathy, the ability to feel another's pain, can serve as a good example.[115] Neural mechanisms at the basis of empathy are those of mirror neurons, playing a crucial role in a more complicated neural circuit;[116] they were first discovered in monkeys. There is, however, reason to believe that other mechanisms that cause emotional contagion, spreading similar feeling from one member of animal group to another, exist in less cognitively advanced species.[117] These mechanisms would be useless and would probably disappear in the course of evolution if not for the huge impact they make on choices: being able to apprehend another's emotion gives an animal a wide range of alternatives of choice, e.g., consolation behavior, which is widespread in chimpanzees,[118] or taking advantage of the sorry state of another individual, which seems to be no less widespread among organisms both non-verbal and endowed with language.

Associative imagination also enables animals to recognize another's consciousness. Non-verbal animals and pre-linguistic children are able to recognize another being's sentient consciousness and desires. This is evident from certain behaviors highlighted by the experiments in the Theory of Mind (ToM) paradigm where human children and mom-human primates change their act following what they can apprehend about other living being's desires, sensory access to stimuli, and knowledge. For example, de Waal quotes research that demonstrated one chimp obtaining an object its older friend tried to obtain and then delivering it to the latter.[119] There are plenty of similar findings in different species. Rhesus macaques have been found to demonstrate apprehension of another's mind by stealing grapes selectively from experimenters who could not see the act rather than from those who obviously (to the macaques) could;[120] ravens were found to approach stores of food established by other ravens when the latter could not watch and even engage in what we could conceptualize as deception;[121] and more.

Recognition of other minds is crucial for developing communications and language. Without it, language would not ever be developed and learned: there simply would be no possibility of such ever occurring to a living being.[122] Language learning and linguistic practice involves communication between different organisms. Yet in order to exchange any sort of communicative content with another entity, an organism needs to possess an expectation that such entity will be receptive to it, i.e., similar to the communicator in the relevant dimension. In other words, it needs to expect another organism to possess a mind like its own. This is a necessary pre-condition not only for language but also for other kinds of communication between animals, e.g., mating displays in birds or displays of aggression in canines: without an expectation for the recognition of an intent by another species member, such displays would not be targeting them, if they would survive the pressures of natural selection at all. Similarly, the sort of advanced social relations we observe in more cognitively developed animals like rats, deer, or monkeys, would be impossible without them recognizing each other as similar enough in their perceptions, desires, and capabilities – in their minds. So preconceptual recognition of other minds has to be in place as a necessary condition for further linguistic development.

Associative imagination seems to provide sufficient resources for the development of such recognition: associating its own mental content with another's behavior, which will also include another's behavior that demonstrates similar consciousness of consciousness.[123]

How would all this enable animal communication? Non-verbal animals' associative mind can connect between different images, and such connections can be quite sophisticated. For example, many animals can connect certain sounds, facial expressions, and body postures to images.[124] This way, they can engage in signaling: producing certain signals, be it sounds or gestures, and understanding their meaning by connecting them to images available for recollection. This also allows teaching animals to understand symbols: for example, it is possible to teach dogs verbal commands. Numerous attempts have been made to teach chimps human sign language. They mastered hundreds of signs and used them in appropriate contexts.[125] However, animal communication systems are principally different from human language: they lack concepts. Rather, they are sets of signals referring to some sensory content.[126] Apes mastering signs could not understand grammar and use word gestures as images associated with other images. One much cited example, allegedly to the contrary, is of a chimp named Washoe who, looking at a swan and being asked "that what?," responded with two signs: "water" and "bird." However, as researchers note, "there is no basis for concluding that Washoe was characterizing the swan as a 'bird that inhabits water.'"[127] Yet even this example, standing in shining aloneness among the attempts to teach language to chimps, non-verbal animals otherwise most advanced cognitively and closest to us genetically, can hardly point to anything but to animal mind coming one bit closer to human conceptual mind at its outmost reaches, yet not quite reaching its level. It is also important to note that in their natural environment apes and other non-human animals do not create anything but signaling systems relying on associations, where some naturally appearing signal, e.g., vocalization associated with fear or facial expression tied to aggression, starts getting apprehended by other members of the animal group as tied to a certain image or complex of images: fear, aggression, etc.

Instincts and Choice

An argument can be made that animal choice is impossible since non-verbal animals are driven by instincts. Therefore, it is important to address the notion of instinct here. The term refers to an innate tendency,[128] or a pattern of behavior an organism gets with its genes. The most common example is nest building in birds. Many bird species, after being exposed to a certain environmental trigger, e.g., change of season, start building quite elaborate nests, where the level of sophistication is such that explicit teaching would be required in humans to achieve similar quality and exactitude. Another common example of instinctual behavior is mating rituals that in many cases involve elaborate dances or decorations – again, something that humans should be explicitly taught. There is no way to explain such behaviors but to concede that many steps of the behavioral sequences in question are somehow biologically encoded in the organisms of species that demonstrate them.

Some instinctual behaviors include elaborate and very exact motor patters. Lorenz describes how a weaver bird (*Quelia*) can perform a highly complex series of movements of tying a blade of grass, even without anything resembling a blade of grass being present; he also provides an example of a goshawk, reared in captivity, who, seeing a pheasant, immediately ambushed it in a characteristically swooping movement and killed it, as if pre-programmed to do so.[129] Dogs shake water off in a fashion common to many, if not all, breeds; salmon migrate many hundreds of miles; baby turtles rush for the sea immediately after they hatch – there are plenty of example of instinctual behaviors that stun us in their exactitude. One is tempted to conclude that instinct prevents choice. Yet on closer examination instinct requires choice.

Instinct in animals defines both some behavioral, mostly motor, sequences, and the goal, e.g., building a nest. Yet it cannot define specific actions. Even with weaver bird's innate skill at tying blades of grass, no instinct could determine that it picks this rather than that blade of grass, or chooses this rather than that tree to build its nest – simply because neither of the particular blades of grass nor this or that particular tree, not to mention cell phone towers, were present when the instinct was developed evolutionary eons ago. The weaver bird has to be equipped with a mechanism to choose its nesting site. The criteria would be instinctually determined, yet this is exactly what would necessitate choice of location, as it will necessitate choice of materials from what is at hand. This way, animals endowed with instincts have to be equipped with the ability to choose between alternatives.

Instinct necessitates choice, yet it also defines its parameters through defining the goal and certain motor activities to be exercised while achieving this goal. These limitations can be tighter or looser. This way, these would create a continuum from an instinct that leaves little choice, and thus limits animal's chances at survival when the environment changes, to instincts that allow more space for freedom, all the way until instincts disappear in humans.[130]

There are also many behaviors, even in animals that demonstrate little cognitive prowess otherwise, that cannot be instinctual, i.e., they do not demonstrate an inherited behavioral pattern. Already in the nineteenth century William James cites experiments that have shown frogs imprisoned in a glass sphere full of water to look for another route to fresh air when the direct one is blocked.[131] Recent evidence for the existence of cognitive maps, the ability to orient themselves in space and change the path in order to get to the source of food or escape danger, in reptiles[132] points to the same phenomenon. Anybody who observed dogs knows that they quickly learn to navigate the house and, seeing a squirrel through the window, know to use the path through the doggy door in order to get to the backyard – even though neither of the steps to do so could have been encoded instinctually.

All these examples point to goal-directed behaviors that do not involve inherited behavioral patterns. Yet none points to conscious deliberation, decision making based on criteria external to specific choices: a dog that elects to assume a supplicant position rather than a snatch-and-run option is certainly not deliberating about the morality of each in achieving the coveted treat. Perhaps, he chooses based on the sensory memories attached to each option, and thus in one family a dog will beg, while in another take a more active stance. In order to make choices based on external criteria that range over multiple, potentially infinite behavioral options, the choosing agent needs the ability to have non-sensory, abstract mental content relate to images represented in its mind. To continue our example with a dog snatching a treat or begging for it, normative considerations of permissible vs. impermissible, sub-species of good vs. bad, is inaccessible to the dog because whichever criterion for goodness we think of, it would be abstract, non-sensory, not something that can be generalized from images to which the agent has been exposed over its personal history. The only mechanism to allow the cognitive operation of normative judgment is language that, as Hegel noted, speaks in universals – and dogs do not possess language.[133] Using such mechanism that operates with universals, we can reflect not only about the alternatives but about anything else, including then mechanism itself – and thus possess knowledge.

Therefore, as far as much of animal behavior is concerned, a certain desired result, condition, or end might be genetically determined or conditioned earlier in animal's life time, yet the behavior itself, as goal-driven, will admit of a teleological rather than mechanistic explanation.[134] As such, it will require choice, either of one specific state of affairs among several that match the desired goal, e.g., three patches of grass the sensory imprints of which our goat has, or to the very least of the means to achieve the goal – the path, the specific steps, etc. In this sense, any goal-directedness, including instinctually determined one, necessitates choice.

Animal Culture and Morality?

While some argue that non-verbal animals cannot be free as they are driven by instincts, others maintain that animals are as developed as humans in terms of their choices, or that human choices are principally no different from the animal ones. Phenomena like culture and justice, that are usually seen as requiring declarative learning and thus language, are sometimes seen as something we share with animals.

In 1967 Stephenson published a paper describing an experiment where monkeys have been trained to avoid certain object by being exposed to strong air current upon approaching it.[135] When placed with other monkeys, naïve of the adverse consequence of approaching the object, the experienced fellows prevented them from touching it by all means available, including physically pulling the newcomers away. After that, male monkeys placed in the cage with the object alone were wary of approaching it; interestingly, females overwhelmingly decided to rely on their own experience and enjoyed playing with the new attraction. This has been hailed as a source for culture: after all many traditions, e.g., dietary rules in some religions or incest prohibition before genetic research, also seem to be as impervious to the experience of subsequent generations as the object avoidance rule to the monkeys. However, it seems that simple conditioning can explain subjects' behavior in Stephenson's and similar experiments: it is the association with an unpleasant consequence that can cause avoidance. Curiosity, a mighty drive in all primates, competes with it, and some participants, most notably Stephenson's female monkeys, chose to go by it rather than by the learned association.

Another famous experiment, by Brosnan and de Waal,[136] is widely interpreted as pointing to a sense of fairness and inequality aversion in monkeys. In the experiment, two capuchin monkeys who could observe each other were both given a cucumber, and both were perfectly content. Then the experimenter gave one of them grapes, understandably preferred by capuchins over cucumbers. When the experimenter gave the other capuchin cucumbers, the latter refused to accept them, demonstrating clear and rather emotional displeasure. This result, however, can be interpreted in terms of frustrated expectation: the capuchin, seeing his fellow getting grapes, comes to expect the same, as now he senses that grapes are available. The gap between the expectation and the reality is what frustrates the monkey. This is very different from the sense of fairness, which, whether we see it in terms of just deserts or in terms of essential equality that calls for equal treatment, requires discursive thinking: the concept of justice to the very least. The behavior of the monkeys receiving grapes, which shows no signs of distress or least care for their less fortunate brethren, shows that the concern is not universal, which would not be the case if they were considering injustice – even though similar behavior is all too common in verbal primates.

The Nature of Animal Choice

The mind of non-verbal animals, as it has been argued earlier, in its higher reaches is characterized by associative imagination that gives animal mind the ability to represent different courses of action, thus enabling choice. The nature of this choice, as well as the areas where it can be applied, are determined by the features of animal mind.

The most important of those features is animal mind's non-conceptual character. Intuited objects and recollected images are *manifest* to the animal, it lacks the capacity to bring them under categories and operate with categories. It cannot theorize, build hypotheses and conceive of different ways to verify them – all what it has are images, intuited or recollected. Such images can be generalized, as it is evident from conditioning where stimuli close in the sensory dimension to the conditioned one trigger the behavior associated with the conditioned stimulus. For example, dogs can be conditioned to respond in one way to a circle, and in another way to an oval. Ovals more similar to a circle would elicit one conditioned response, and ovals slightly more circular than the original one would elicit another; exposure to a shape in-between would cause adverse behavioral reaction deemed "experimental neurosis." Similar effects have been found in other animals, e.g., sheep.[137] However, generalizations are still images, they re-present sensible content extracted from the originally intuited images.[138] The only way for intuition to be associated with them is to have common sensible content. Generalizations are not abstract categories like number, good, humanity, evolution, philosophical thinking, or free choice: these go beyond images, and generalization available to animal mind are always sensory. Animals also cannot construct new generalizations at their own will, they are limited by the sensory experiences they've had and the ability to associate between them that is determined genetically. Some animals can go beyond conditioning and build cognitive maps, yet this ability does not progress farther than associating generalized content either: cognitive maps are devices to place images in a certain relation to each other, a relation that reflects their location in space. Unlike human maps where agreed-upon signs stand for objects in the spatial domain and preserve the relevant aspect of their geographical relation, animal cognitive maps are associations of images as they are related to each other in space, there is no signification or any other sport of conscious abstracting of features there, no *stands for* relationship. Consequently, while non-verbal animals can demonstrate quite remarkable competence in various areas, it is, as Dennett puts it, "competence without comprehension,"[139] competence that cannot go meta on itself, and this severely limits choice.

Animals are restricted to choosing between the courses of action suggested by the images they intuit and have in store. In this sense, non-verbal animals are *poor in world* (*weltarm*): they are severely restricted by their immediate relationship with the series of elements to which their senses are receptive, whether by intuition or by recollection.[140] The nature of these relations is defined and cannot be changed by non-verbal animals: they lack the ability to abstract, to apprehend something *as* something, to look at their ways of acquiring information *as* ways of acquiring information, their ways to associate images as ways to associate images, and thus conceive of the possibility of altering the existing ways or introducing new ones; "the animal remains enclosed in the circle of its environment and can never open itself into a world."[141] In contrast to a human world of which human beings are conscious co-creators, animals have an environment of which they can never make conceptual sense (in terms of composition, type, fairness, suitability, etc.) nor plan to change it and carry out the plan while comparing their actions and results to what had been outlined.

The playground of freedom in animals is the space around them and the time plane: past sensory impressions, which can be generalized yet still remain images, and the future toward which the choice is pointing. Yet the choice itself is between the alternatives imagined in the literal sense of this word, imagined based on stored sensory content. There are no plans to be made; no abstract, non-sensible criteria to apply to different options – no non-sensible plane at all. Some animals are capable of empathy and can choose between the course suggested by empathy and fear, for example – yet there will be no weighing of merits in accordance to any moral principle, be it utility or universality: non-verbal animals lack the cognitive ability to apprehend abstract principles. Non-verbal consciousness is necessarily characterized by unquestioning attitude: pre-discursive mind is not able to question things, as it cannot evaluate alternatives in light of consciously selected external criteria – this requires linguistic intelligence. Decisions whether pleasure is preferable to loyalty, how should an agent decide what is the acceptable level of risk, and whether fear should be suppressed for the sake of conviction in the rightness of action all require contemplating criteria for decision making. Criteria of choice cannot be selected based on generalization of sensory images: they are abstract and address states of affairs based on their belonging to non-sensory categories, e.g., moral or immoral, useful or not useful, or beautiful or ugly. Therefore, animals cannot engage in self-improvement or world-changing projects. Animal choice thus is quite limited both in terms of the scope of available alternatives and severely in terms of the dimensions of choice.

Similar type of choice takes place in non-verbal humans, e.g., pre-linguistic children or adults who, due to a neural deficiency, do not possess language. Both groups have access only to the associative imagination and thus are incapable of conceptual choice.[142] It is important to emphasize that neural deficiency here pertains to the capacity to abstract from image content and thus to mind's ability to reflect upon its own workings, and not to the capacity to produce vocalizations or any other forms of language expression: a living being in possession of conceptual mind is perfectly capable to choose between alternatives based on their conceptual features rather than on their sensory aspects alone.

With all the limitations of animal choice, it is a self-determined choice between alternatives nevertheless. Animal organism can apprehend alternatives, choose between them, and carry out the course of action associated with the chosen option. Therefore, it makes sense to talk about freedom in the context of animals.

HUMANS: RATIONAL FREEDOM

Human and Other Animals: Behavioral Difference and Physiological Similarity

Humans exhibit a range of activities that are very different from any other living being both in terms of scope and in the way they are carried out. A small subset of human actions looks very similar to what is characteristic of all life: coming to be, nutrition, and death are the most prominent – the Aristotelean nutritive soul persists in humans like in all life. A bit wider set is similar to what other animals, particularly primates, do: mating, child rearing, engaging with new objects, articulating emotions vocally and by means of facial expressions, etc. However, a deeper examination shows that very few of our behaviors are similar in terms of their mental and social context to what our closest relatives, the great apes, engage in. The way we rear our children is subject to traditions passed on orally and to intense debate where much of the argumentation relies on the findings of social and other sciences. People do not mate but enter into relationships, which are also subject to various legal, customary, and assumed rules which we discuss, change, and at times consciously break for a variety of reasons. We do not just engage with new objects, we discover them: contemplate their novelty, try to categorize them, discuss our findings, and receive rewards for our discoveries. Our emotional life is subject to much scrutiny: we articulate and discuss principles pertaining to the permissibility of expressing emotions, the appropriateness of emotional expressions in specific settings, and more. We think, deliberate, discuss things verbally and in writing – we imbue objects and actions with meaning, create new objects and contexts, propose symbols and signs, build and bomb cities, and are ready to change the courses of our lives, sometimes dramatically, in light of the results of these deliberations. No other animal does anything similar. We do not only do much more than other animals, we also do it very differently.

With that, we and our primate relatives have remarkably similar genetic endowment[143] and, consequently, physiology. This pertains to the kind and arrangement of bodily organs, central and peripheral neural system, metabolism, and the types of motility we can exercise without technology. The main differences seem to be not even the vocal cords and their coordination, as it had been claimed for decades, but the brain: its size and composition.[144] The average human brain weighs some 3.5 time more than a chimp's: 1,352g vs. 384g; it also has more white matter in the temporal cortex, which points to more connections between nerve cells and thus suggests greater computational power, or ability to process data.[145] The sheer weight of the brain seems highly problematic as a crucial difference: after all, blue whales have much heavier brains. The second difference, the amount of white matter which is assumed to lead to a greater processing power, is merely quantitative; it is also possible that this is a result of nurture rather than nature: a human brain continues to grow long past the time when a chimp's reaches its full mass.

The apes, if we consider them in terms of the aspects of mind addressed earlier, are also remarkably similar to us: higher mammals and humans are similarly capable of intuiting, storing, and retrieving images; generalization over sensory aspects of stored images; producing and interpreting communicative signals – both cry when distressed, scream when angry, and laugh when happy; it is not too hard for people and dogs, for example, to interpret facial and vocal expressions of each other's emotions. In fact, many animals can be trained to follow commands, thus associating sensory images with certain behaviors.

There is, though, one crucial difference between humans and other animals that accounts for the behavioral differences pointed above – human language.

Language and Discursive Metacognition

The Basic Property and Merge: Recursive and Reflexive

Non-verbal animals develop extensive signaling systems. They also can be taught to react in specific ways to words in human language or images or even use hundreds of signs in a human sign language, as the cases of chimpanzees Nim,[146] Washoe,[147] and others[148] demonstrated. Yet human language amounts to something principally different – it is reflexive, i.e., it can address itself.

Language enables the mind to look at itself by considering the relations it forms as objects, where such consideration is known to the mind, where the mind is conscious of it.[149] Non-verbal animals' mind can relate, or associate images based on generalizing their sensory features, yet this is where it stops. It can have as many such relations as its computational and storage capacity allows, even dilute certain aspects of images in favor of others, but this web of relations is flat. Non-verbal animals can relate images to one another or to a generalized representation based on some sort of sensory resemblance; in a similar way they can recall a memory following an intuition that shares a sensory aspect with it. Yet there is no level of cognition in non-verbal animals that would enable consideration of the *relation* between images as opposed to images themselves, consideration not of the images it stores and recalls but of the relations among them as its object. Such consideration would be free of sensory content, as it focuses not on the sensed aspects but on the relation between their bearers. This is why this type of consideration would also enable relating representations to imageless universals – it is not limited to imagery any longer.

The consideration of images by the minds of non-verbal animals, as noted earlier, involves relating to them through feeling. For a dog, recalling an image of a significant human involves pleasure, as looking in the eyes of a human is associated with the release of oxytocin that is highly pleasurable to dogs. The same association applies to humans.[150] This, together with other experiences that bring pleasure, can certainly develop into a feeling of love toward a specific human – and then encountering this human will bring joy by association. However, to ask questions *about* the relation between catching the gaze of the significant other and the felt pleasure one needs more than the association itself. A dog's love, which has been lauded so many times in literature and by any person who has dogs in his life, is absolute in the sense that the dog does not question it – it is pure feeling. Dogs cannot question it because they lack the cognitive resources to consider the relation of love itself, to develop mental contents *about* this love. This ability is achieved through language.[151] This is what enables humans to ask about love "What's in it for me?" – but also questions like, "What does it mean 'to love somebody'?" and "What are my duties to the loved ones?" – questions that make human ethical love principally different from animal love.

This sort of aboutness cannot be achieved by generalization. It requires a look from above, to use a spatial metaphor, or a capacity to go *meta*, in our case – mind's capacity to address the non-sensory contents of its own making, i.e., relations between remembered images and the relations between those relations, the only limit being the capacity of our memory and the processing capabilities of our central nervous system. Such capacity is usually deemed *metacognition*. The clearest example here would be, perhaps, that of normative consideration. To ask "is it good to love?"[152] one needs not to experience an intense feeling of pleasure at the recollection of somebody but to relate this sort of relation between the self and the other to something else, a criterion that would be external to the feeling itself. As such, this criterion would be abstract, non-sensory, for example, Kant's Categorical Imperative or Bentham's Utility Principle. Yet those principles themselves are subject to the same sort of scrutiny – they can be questioned in terms of coherence, applicability, etc.; in fact, they frequently are being questioned. Even when the consideration of certain behavior is decidedly short, and the answer is not suggestive of deep analysis, we have a similar situation. For example, a child might ask regarding a certain ritual or dietary restriction "Why do we do this?" – and get a curt "Because this is how we do it!" answer. Yet the ability of the child to ask why and the fact that the parent understands the question perfectly well discloses the aboutness of language, our capacity and propensity to go meta, or metacognition. As we will see soon, this has crucial impact on human freedom.

For metacognition to have an impact, the mind needs cognitive tools that enable it to consider any content, sensory and non-sensory alike; to expand its purview beyond the sensory – to think. Thinking will be addressed in the next section.[153] Now we need to consider what would such tools be.

When the mind acquires the ability to consider associations between encountered and remembered images, to look at it as it looks at the related images themselves, it gains the ability to form a hierarchical system of meaning where elements not only connect to each other but also range one over the others, as in verbal categories. These multi-level systems of meaning are what we observe in human languages yet not in animal communication systems – in fact, this very sentence presents such hierarchy, where the terms 'human' and 'animal' are not merely generalizations of images but categories that range over any possible human being and non-verbal animal. This is what Chomsky deems the Basic Property of language:

> each language provides an unbounded array of hierarchically structured expressions that receive interpretations at two interfaces, sensorimotor for externalization and conceptual-intentional for mental processes.[154]

Chomsky suggests that the possibility of constructing such unbounded hierarchical system relies on one operation that non-verbal animals lack yet humans possess – the operation he calls Merge.[155] Merge is the ability to take two existing elements, e.g., images, and combine them together, thus creating a new image while preserving the original ones. This way, we can take a word 'drum' and a word 'beat,' both probably of an echomimetic origin and thus symbols connected to the image it represents in terms of auditory similarity,[156] and, using Merge, connect them into a word 'drumbeat' that would have a new meaning, referring to the beat of a drum. The same word can later be used metaphorically, as in 'deafening patriotic drumbeat' – the same Merge operation, working with the existing elements of language, is responsible for joining 'drumbeat,' a result of first level of abstraction, a more abstract 'deafening,' and one of the more abstract and shape-shifting words in human languages, 'patriotism,' a result of many prior Merges and undoubtedly subject to many future ones.

The general nature of the Merge operation is also at the root of another feature of language that plays a formative role in human freedom – linguistic reflexivity. Merge can take any inputs, apply to any content, including the outcome of previous merge operations. This way, linguistic mind can relate to itself, describe its own contents and workings using the very same resources it utilizes in describing images. In animal signaling systems distinct expressions, vocal or other, are connected to certain psychological states: fear, hunger, lust, aggression, etc.[157] These signals can be quite sophisticated and use multiple sensory modalities. They are subject to positive reinforcement or suppression by the response animal organism encounters: a weaker male who emits too loud of a signal to indicate his desire to mate might well face aggression from stronger males, a response that is likely to condition him to be quieter next time. However, animal signals only convey psychological states, and thus are merely outgoing. With Merge at their disposal, human mind can not only construct more complex messages, produce informational content that reflects multiple aspects of psychological status, as in "pain in the leg." Crucially, it also allows addressing linguistic content itself, as in "it is very unfortunate that the left leg is in such an excruciating pain." In this example, the Merge first constructed a message from "pain" and "leg"; then it addressed this very construct by evaluating it – "excruciating"; then it addressed the "excruciating pain in the left leg" by evaluating it as "unfortunate." This repeated application of Merge to its own products opens a new dimension of meaning by enabling the mind to reflect on anything – not only the outputs of the senses and felt emotional states but also abstract concepts. In fact, addressing own content is the first step toward abstraction. Putting together evaluative criteria, conceptualizing emotions and cognitions as separate, morality – all these are made possible by the reflexivity of the mind accorded by Merge.

It is important to note that language in general and Merge property in particular refer to what Chomsky deems the *I-language*, where "I" designates internal, individual, and intentional.[158] It refers to a computational procedure, not to a set of objects it generates. The latter is usually deemed E-language – an external language.[159]

As it is evident from the examples, Merge works recursively, it can be applied repeatedly to its own results. There is no limit on the number of times Merge is applied. Thus, there is no principal limit to the hierarchically structured expressions, to the number of layers of the language web – to the size of the vocabulary and the repertoire of meanings words and sentences express. The limits are practical: the storage capacity of the human brain which will determine the size of the vocabulary used by an individual. The recursive Merge enables the reflexivity of the mind: language is our means to address anything, including language itself. By using association, mind can make a cognitive map of its environment and later use it to navigate the environment, mostly in a literal sense: rats or, as some research argues, even bees do that.[160] By using language, mind can try and cognize itself,[161] build models of itself, and even develop and consider new paths for own development.

This leads us back to metacognition. The term itself, when used to address the ability of mind to relate to its own contents without limitation, and specifically to the contents that do not have a sensory component, sound ambiguous: after all, metacognition, or cognition of cognitions, is a characteristic of any mind. Intuited and remembered images, as well as their generalizations, are as much mind's own contents as the relations between them, ethical principles or grammar rules; relating to them is relating to mind's own contents. Since it is language that enables a different kind of metacognition that is described above, I will from now on refer to the ability to address imageless as well as figurative mental content, as *discursive metacognition.*

From Associative Imagination to Human Language

As it is the case with any new life form, the human one utilizes the resources available to the non-verbal animal life form and starts using them differently. In the case of language, this process takes place as well.

Some non-verbal animals, e.g., great apes, have the ability to form general representations of images, associate them, and retrieve them and associated images following a clue.[162] The next step down the road toward the development of language is semiotic imagination – "rendering these aspects the figurative expression of that general representation."[163] The difference from association is subtle yet crucial: here we have an image that is not merely associated with other images but *stands for* other images. What is crucial for this task is reflexivity, the ability to consider own mental content as an object. This is the same ability that stands behind Chomsky's Merge, where two elements, units of meaning in mind are combined to produce a third one. The *stands for* relation abstracts the association in having it linked with particular aspects of some images, yet not associated with specific images: any image possessing such an aspect can be represented by the image chosen to stand for others, thus laying the foundations of linguistic universality.[164]

The most elementary form of the *stands for* relation is symbolization. Symbols resemble what they stand for, or what they symbolize, by sharing one or more features with it.[165] With simpler symbols, the resemblance is sensory: for example, red can stand for blood and blue for sky. Yet, when equipped with the reflexive ability, human minds can continue the symbolization process by further abstraction, going meta, considering the relation just formed and its links to other relations. Red now can stand for fighting and blue for open, as fighting leads to bloodshed and sky is associated with open spaces. This process can continue further, involving abstract entities that appear at the stage of fully developed language. Here, red would stand for struggle, as it frequently involves fighting, and blue for freedom, which is associated with the lack of confines – like in open spaces. Symbols this way are not only a necessary pre-condition for the development of language but persist in it: flags and ranks, no less prominent today than in times past, attest to that.

Many non-verbal animals can comprehend symbols. As symbols physically resemble what they stand for, all what is required here is the capacity for general representation, the same ability that is at work in associating images. The ability of non-verbal animals to associate different sensations has been observed since times immemorial, including in controlled scientific settings; monkeys were found to possess even such advanced capacity of symbolic thinking as associating tokens with foods and demonstrating preferences in choosing and exchanging tokens that would lead to the maximization of pleasure.[166] From here, there is nothing that precludes production of symbols by non-verbal animals: all what is required here is associating externalized general representation with this representation as a mental content.[167] Research, however, finds little evidence of symbol production in animals.[168] The reason might be rather simple: we are looking for symbols like our own, first and foremost – those that are expressed by pictorial images. It is hard to find in the animal kingdom creatures whose motor apparatus would be a good fit for drawing pictures and who would have the requisite materials available. To discover production of symbols in animals, we might look at other forms of expression. For example, apes and canines are known for displays that indicate aggressive intent.[169] Initially a particular expression of aggression may be related to the physical readiness to attack.[170] However, the association between an animal's own bodily state accessible through proprioception, own feeling, and the response of others might well make dogs, cats, apes, and other animals externalize the appearance of such readiness in order to scare the opponent: all that is required here is the capacity for association and externalization of generalized representation. This would be a rudimentary symbol, similar to humans intentionally "making a face" in order to signal sadness or raising a hand with a palm turned forward to signal another to stop. The capacity to choose between alternatives would also be at work here: it is within the capabilities of a pre-verbal animal endowed with associative imagination to envision possible outcomes of baring its teeth or lowering its head.

Yet even if we take into consideration these possible cases of production of symbols in non-verbal animals, we would have to admit that such activity is much more rare in them than it is in humans; and that the symbols themselves do not stretch beyond what can be closely associated with what is symbolized on some physically accessible scale: the distance between baring one's teeth intending to scare the opponent and doing the same as part of a genuine intent to attack is much smaller than the one between shedding blood and a small red band on the uniform of somebody who demonstrated courage in battle. This is because non-verbal animals cannot produce a symbol with an intent to produce the symbol. Such intent would require an understanding of the symbolic relation, the relation of one image *standing for* another. This necessitates discursive metacognition.

Once the agent understands the nature of the symbolic relation, production of symbols becomes prolific. Discursive agents now do not have to experience the reaction of their communication targets to the relation between the symbol and the symbolized. Instead, they look for opportunities to use images that are easier to communicate as standing for other images, those that are hard to communicate, e.g., images of things that are located elsewhere at the time of communication. For example, three outlines of a cow can stand for three cows. Consequently, this understanding of the symbolic relation as the relation of *standing for* leads to the realization that images do not have to share any sensory aspect in order to stand one for another – it can be *agreed* that X stands for Y, and the purpose of communicating meaning will be achieved without any sensory resemblance between X and Y.

Symbols alone cannot account for language, as they still bear resemblance to mental content in some sensory dimension, be it visual, auditory, or other. Consequently, symbolization alone does not allow for abstract mental content that is not grounded in once-intuited imagery. However, symbols, by having their connections to what is symbolized as a relation of *standing for* rather than *being associated with*, constitute a crucial step toward loosening and then abandoning altogether the sensory connection between mental content that stands for something and this something it stands for.

This gradual sensory disengagement of the symbol from what it symbolizes leads us to signs – intuited content that has no sensory, intrinsic connection to what is stands for.[171] The link drawn between the sign and what it stands for is arbitrary, initially subjective. It expands the range of referential activity: now a mind can alter the connects, expand or shrink the range of things to which a sign refers with no dependence on the sensory characteristics of the sign. This is where, for example, three horizontal lines can stand for a food storage, and three vertical ones for a weapons' cache; a certain utterance, e.g., the sound combination 'cow,' can stand for a type of objects – namely, cows, to borrow and example from Vygotsky,[172] without having anything to do with the sensory aspects of perceiving a cow; etc. Such sign, however, is still pre-linguistic: its meaning is still tied to intuited content, both of the sign and of the signified.

The next step toward language is naming. Reflexive mind here apprehends the connection between the sign and the general representation for which it stands,[173] and the name becomes a representation of such relationship. As the phylogenesis of language is repeated by its ontogenesis, this step can be exemplified by Vygotsky's observation of the difficulty children have in applying a different verbal sign to what they know to be tied to another word: when asked to use 'cow' instead of 'dog,' i.e., to use the word 'cow' to refer to dogs, his young subjects had trouble answering correctly the question whether dogs have horns, and claimed that they would have smaller ones than real cows.[174] The essence of the name as opposed to sign is understanding the arbitrary nature of signification, the severance of the vestiges of sensory connection. The arbitrary nature of naming leads to what is perhaps the greatest expansion of the scope of language: now being endowed with language "can name what isn't there," as Knausgaard puts it:[175] events and places one never sees but reads about, people who are not among the living, even concepts like silence. This way, naming provides the basis for the "tremendous expansion of the lifeworld that occurs through language"[176] – and with it of the scope of freedom.

No language consists in naming alone – names are merely the units combined by language into propositions according to rules. Only this way conceptual content can be communicated: truth and falsity, values, conventions, arguments, etc. Language has not only content aboutness, having nouns refer to general representations and to abstract concepts with no intuited reference, but also structural aboutness – the ability to address material and conceptual relations by structuring relations between words.

Verbal memory provides a bridge to such language.[177] Firstly, it can retain the name itself, the connection between the sign and the intuition, and retrieve the corresponding meaning upon encountering a physical configuration of a sign. This enables mind to associate names themselves. Here the mind operates without anything immediately given, any intuited content as a trigger. This opens the door to conceptual thinking: we start thinking in names, and this prepares us to work with "products of intelligence that lack image content," or the concepts.[178] The last step is mechanical verbal memory, or rote memorization, where names are recalled without any connection to their meaning. Not only rote memorization can exemplify mechanical verbal memory: rhyming, which seems to be characteristic of any known culture, can work just as well. This is crucial: the intelligence here gives its own ordering to names, thus combining subjectivity, the order of its own creation, and names, which in this case it treats like objects. The mind here is for the first time engages in creating its own order of things that are used not as intuited, acting as a mental *causa sui* – it is no more empty without the intuitions.[179]

The unity of subjectivity and objectivity introduced by mechanical verbal memory, a mind that treats its own creations as objects while understanding their nature as its own creations, is what counts as reason *in and for itself* or, as I will try to establish later, rationality. In intuition, the unity of reason was immediately given, as the immediately given object was triggering its intuition by affecting the senses and neural circuity – unity *in itself*. In representation, it was a product of imagination that produced images about something given to the mind, unity *for itself* available as such to the intelligence. With verbal memory, the unity is available to the intelligence as its own activity, thus existing *in and for* intelligence.[180] This prepares the ground for thinking of thinking, where the subject and object are one,[181] to asking questions like "What is the basis for X?," "Am I right?," "Is [proposition that I have in mind] is correct?," but also "What is truth?," and "Can there be truth?" – conceptual questions of meaning and value, considerations of concepts that are in the mind.

The process described above whereby signs and names are formed takes into accounts the capabilities of an individual mind. In reality, though, it is highly unlikely that an individual isolated from a community would develop a significant store of names – it is hard, in fact, to imagine a human individual completely isolated from conspecific community due to the prolonged child rearing in *Homo Sapiens* and the collectivist nature of our primate relatives; it is also hard to imagine what impetus an isolated individual would have for developing a system of names, and how a large system of names, even remotely comparable with the number of nouns in a typical human language, would be created and retained by a lone individual with her limited life experience. Yet most importantly, rules of grammar which give language its structural dimension and expand its capacity to enable an infinite number of expressions develop as a result of its communicative function.

The emergence of language is associated with a plurality of individuals.[182] Sharing a similar experience of a common world, due to similar environmental conditions, sensory system, and neurological structures; endowed with the capacity for signification and naming, as well as verbal memory – human individuals leverage these capabilities to communicate in the context of multiple activities in which they are involved – language becomes a language game, communicative activity bound together with rules[183] and changing together with rules as a matter of use.

Rules are quite special. Firstly, in themselves they are a further abstraction – rules have no image content. Secondly, syntactic rules enable using names in a universal fashion: they make names refer to universal categories under which past, present, and future particulars that share relevant features can be subsumed. "Jane is a carpenter" subsumes Jane under an abstract universal category of carpenters that admits a potentially infinite number of individuals. Thirdly, abstract language rules create possibilities for forming new abstract relations. "This is my hatchet" expresses a relation of ownership, an abstract relation that has no image content. It is expressed by virtue of rules that make names with concrete reference, 'I' and 'hatchet,' appear in a way of such relation. For that, new parts of speech that modify nouns or declensions are introduced. From here traveling the road toward higher level of abstraction, e.g., normative language of right, true, and beautiful, is a matter of using the reflexive Merge mechanism recursively.

Language equips agents to consider a dimension crucial for freedom – the temporal one. Non-verbal animals have a degree of temporal thinking, as mentioned earlier, yet they lack the ability to consider temporality as a type of relation between events, to think about time. As a result, they can act with a view of the future yet not plan for the future: planning requires considering future *qua* future, a temporal state that is yet to happen but will happen and, as far as the way of the world is concerned, will bear the same characteristics that the present and the past possess.[184] As Bischof and Köhler hypothesized[185] at the beginning of the twentieth century, non-verbal animals can imagine a future event yet cannot imagine future needs. This was their interpretation of an experiment where chimpanzees would take a stick to get bananas from a farther ground yet not water to drink once they get thirsty on the way if they are up for a trip. So far researchers failed to demonstrate empirical findings that would contradict this conclusion.[186] In order to anticipate the necessity of taking a stick, one needs to imagine the situation of removing bananas from a farther ground, or to recall it from memory if such ground with the coveted fruit had been encountered already, which was the case with the original experiment. This situation is easy to imagine when a chimp feels hungry: after all, bananas are closely associated with rectifying hunger, and using a stick to get them associated with the bananas in the context of the place where they have been found last time. There is no need to grasp the differences between the past, present, and future for this mental feat. Yet to consider future need – our case, thirst that will be felt after traveling in the sun for a while – it is not enough to imagine oneself experiencing it. This requires considering how needs arise over time and relating this process to the conditions that cause the need and its satisfaction. In other words, it requires considering the progression of time going from the present into the future and changes in the strength of the need in question along this line. Imagining states of affairs, including states of one's own sensation, would not help here: the arrow of time cannot be imagined this way. In fact, it cannot be imagined at all: it does not have a sensory component to it. Since it does not have a sensory component, any mental operation that involves relating sensations to the arrow of time requires cognitive tools that would "apprehend universal contents irreducible to any image,"[187] or thinking. This, in turn, requires signs that are inaccessible to non-verbal animals, since they, differently from symbols, have no intrinsic connection to what they stand for.[188] Thus, no wonder that apes can act upon their present needs yet not upon the future ones that are yet to be felt. In fact, this is not always easy for humans as well – anybody who has ever forgotten for the third time an essential item when going on a hike can attest to that. Psychological research suggests that the ability to plan for future needs appears in children around age four,[189] the age crucial for the development of language.

It is the semantic simulation, fostered by the rules of our language games, that enables the "construction of a detailed mental representation of a general or abstract state of the world"[190] – conceptual, ranging over many and potentially infinite number of instances. Without that, planning one's life or a project, putting forward a hypothesis and testing it, considering changing the world or consequences of a possible change are impossible. All these are included in our range of choices thanks to linguistic intelligence.

Rational Intelligence

Aristotle's *zoon logon echon* (ζῷον λόγον ἔχον)[191] is usually translated as *rational animal*, or *animal that has reason*. However, notes Taylor, *logos*, the word translated here as *rationale* or *reason*, "allows a full stretch of polysemy": in different contexts it can also mean 'word,' 'discourse,' and 'account.'[192] Amartya Sen, in his search for a definition of rationality that would apply to different aspects of human endeavor, from morality to economics, interprets rationality "as the discipline of subjecting one's choices – of actions as well as of objectives, values, and priorities – to reasoned scrutiny."[193] This definition is intentionally broad, as rationality pervades all aspects of human enterprise, yet the main motivation behind such interpretation for Sen is avoiding arbitrary assumptions, foundations that cannot be justified, e.g., "rationality is an enlightened self-interest" or "rationality is internal consistency."[194] Indeed, conceiving of rationality as reason's scrutiny of agent's choices, or self-scrutiny of a human being, makes it found itself yet at the same time captures the defining characteristic of human mind: its conscious reflexivity. Language is the vehicle of such reflexivity. It enables verbal living beings to represent possibilities as alternative choices, including providing a rich description of each choice in context together with its temporal aspect: historical, rooted in the past; contemporary, as relating to the present; and future-oriented, in terms of steps required and possible consequences. Language is the connection between the different meanings of *logos* in the context of humans, logical/discursive animals: language is necessary for rational life, as reason consists in giving account of things, including an account of itself, and language is the means of articulating such an account, thinking about it, and communicating it to others.

Pre-conscious psyche of simpler animals is capable of expressing feelings and developing habits, as it was described earlier. Conscious animals, capable of distinguishing between subject and object, are capable of much more: forming and retaining images of sensory impressions and treating them *qua* images, i.e., as something different form intuitions, without being aware of the distinction itself; associating images and generalizing over them based on sensory dimensions, without being aware of the nature of the relation; recognizing another's consciousness without conceptualizing it as consciousness. As such, animal consciousness is turned toward the world: it focuses on the present moment with the ability to leverage past experiences and envision the future only as it relates to the present: by recollecting past experiences and considering future actions only as they pertain to present needs and environmental triggers. Its experience of selfhood is phenomenological, its awareness of the self does not go beyond the level of separating between subject and object.[195]

With language, rational human consciousness is born. Non-verbal animals endowed with mind possess self-consciousness, as they are necessarily aware of themselves as distinct from their Umwelt: otherwise they would not be able to function – move toward food, avoid dangers, mate, etc.[196] Yet a non-verbal animal, lacking discursive metacognition, cannot form the concept of the self that depends upon language, cannot conceive of itself as a self. Human self-consciousness, on the other hand, is a self known to itself as a self. By becoming the object of its own mind, a self can be put under general representations and family resemblances – this is a feat accessible to pre-linguistic consciousness through associating sensory aspects of images, including images of own feelings and proprioceptively accessed images of own bodily states. Yet to relate to the cognition of a self having this or that feature, to think *about* the self having this or that feature, the self should be able to go meta, should have discursive metacognition. To issue a warning cry when being threatened by an animal perceived as a manageable threat rather than to run back to the burrow in response to spotting another animal that registers as more threatening, an organism does not need more than being able to associate sensory aspects of size, danger, and warning cries, etc., that is, very physical aspects of life, and using cognitive maps, perhaps. But to consider whether warning cries in general are a proper response, more is required – the ability to cognitively address the relation between self and warning cries and to relate this relation to an external criterion. Again, normative thinking is a proper example. A non-verbal animal cannot consider warning cries as good or bad, as it cannot form conceptions of good or bad – conceptions that do not have sensory contents yet range over potentially infinite number of images. Linguistically intelligent organisms, on the other hand, are perfectly capable of that. We can address our own relations with our responses and memories, using mental contents as categories inclusive of other mental contents, be it sensory memories or other categories.

Self-conceptualization by means of language can give rise to multiple, potentially infinite ways to understand oneself: as an animal driven by instincts or as a rational being, as a member of an ethnicity or adherent of certain religious creed, a mother, a professional, a failure or a success – or any combination of the above. Language allows us to use categories we acquire in communication with others, as many as there are available; or construct our own. In fact, language accords us the ability to construct a possible self, to form a conception of our own selves as goals or as something to avoid, what has been called in literature Self-Defining Future Projection.[197] This way, language is giving us the account of the self, its *logos*.

Language also provides the means of realizing the aspect of rationality Sen is concerned about – that of providing a disciplined scrutiny of one's own choices. With universal categories, language enables addressing one's own consciousness and other consciousnesses as having the same universal aspect, thus unifying the subjective and the objective.[198] From here, discursive mind can apply the same criteria to itself as it applies to other minds, leading to a disciplined evaluative approach of own choices – a rational approach.

The core of this approach is the conception of truth – "the unity of concept and objectivity,"[199] where objects considered by mind are seen as falling or not falling under universal concepts. Without that, one cannot think of scrutinizing one's own choices in any disciplined fashion: lack of universal criteria independent of the particular content of the self will make any self-consideration a matter of feeling, desire, or arbitrary preference. This is the way by which rationality "knows its own subjective workings to be inherently objective and open to truth and, in that respect, universal to all rational subjects."[200]

A new life form transforms the whole of living being's experience by making different use of its *zoe*, thus creating a new *bios*. Rational linguistic intelligence is the life form of humanity in this exact sense. Differently from the non-verbal animals whose world consists in their sensory experiences stored as more or less generalized image representations, instinctual endowments, and present environmental stimuli, human world is that of culture and normativity. By being discursive, rational beings, humans impose on their sensory experience a new, conceptual character.[201] We construct our selves, as Madhyamika Buddhist philosophers argue: we do not have any self in terms of a persisting conceptually self-aware mental entity besides the one we construct.[202] We are the ones to impose a "unity and coherence on a complex, multifaceted stream of events and processes"[203] that non-verbal animals and pre-verbal children merely experience as their own. Humans know themselves as having a conceptually determined identity that possesses universality. We assign meaning and theoretical significance to what we perceive – we categorize what we sense under universal categories of our own making. We assign truth values and other universal evaluative marks to propositions, i.e., put them under most abstract categories. Linguistic rationality also expands the range of our experience, adding to it thinking thoughts, conceptual contents that are retained by memory and can be retrieved by a conceptual connection rather than by a general sensory property[204] – the associative mechanism of content storage and retrieval is now at work not only with images but also with abstract, imageless contents.

Another dimension of discursive rationality crucial to understanding the character of human freedom is that of normativity. Normativity is a practical, guiding aspect of rationality: concepts not only reflect our experiences but also motivate our actions. Categorizing ourselves as members of communities to which normative significance is attached can tell us what to do in this or that situation. Considering oneself a member of a family, nation, ethnicity, religion, class, political party, scientific community, military force, resistance movement leads to certain actions. Moreover, we can construct new normative communities and create new norms – it is language that enables us to conceive of such communities and norms, communicate them to others, evaluate them, etc. Yet normativity goes beyond communities. Concepts like ethical behavior, human rights, property, duty, religion, aesthetics all have normative significance – and we can create more; we actually do. All this has direct bearing on human freedom.

The Nature of Human Freedom

Like animal choice, human freedom consists in choosing between alternatives. The character of choice, however, is radically different, which brings about a dramatic expansion of the range and type of choices human being can make, including choices to deliberately suppress one's own freedom.

In any situation where a choice is to be made, discursive agent has a richer conception of what can be done than a being that is capable of choice yet is limited to what the association of intuitions and recalled sensory contents can offer. This is because discursive intelligence conceptualizes the choices available, brings them under categories, and then analyzes the alternatives not only by sensory association but also, if not primarily, by the conceptual connections it can make. This has several important consequences.

Firstly, it can generate new alternatives: the more connections one makes, the likelier it is that other options will come into mind's view. As concepts range over sensory memories and other concepts, they form a hierarchy. Each step up the hierarchy potentially brings many more links to other members than a flat web of associations can muster.

Secondly, conceptual thinking can bring about new alternatives that would not be available through sensory association alone, even generalized. Doing nothing as an alternative to consider in a situation of seemingly binary choice between two actions, each of which answers some urgent emotional need, is but the simplest example. One can also think about technological solutions to the limitations of human body, e.g., in terms of it being slow to travel, pass messages over large distance, or preserve information; about convincing others to change their ways – something that can be accessible only to beings endowed with discursive intelligence; or about embarking on a project of changing the parameters of the situation in which the agent in question needs to make choices, e.g., changing the rules according to which the society allocates its resources, thus constructing the opportunities individuals have in several important domains.

Thirdly, the criteria by which discursive beings decide are of a conceptual character. Non-verbal animals do not apply criteria in choosing among alternatives – such application requires conceptual consideration; instead, their criteria work directly, immediately, they are not conscious *of* the criteria they apply. A decision to refrain from barking because it upsets the significant human in a dog's life is likely made following the sensory image of displeasure the dog experienced when its human was unhappy; a decision to follow a command – based on the association of doing so with the sensory image of such human being pleased.[205] Conceptual human mind brings choice alternatives under abstract sets of criteria: those of normativity, authenticity, etc. Yet human mind can go farther: it can choose which criteria to apply to choice and to which criteria to give preference. It can decide to allow itself acting "on a whim," adhere to what it conceives of universal moral standards, or go with what it thinks to be the expectations of the group to which it considers itself to belong. Humans, by virtue of conceptual mind, have the ability to demonstrate questioning attitude: exercise meaningful doubt not only regarding this or that choice yet pertaining to the criteria of choice, or even the means whereby such criteria can be evaluated – this way, potentially *ad infinitum* up the conceptual hierarchy.

Human freedom, again thanks to our conceptual mind, is not limited to making choices in specific situations. Our conceptually reflexive mind enables us to choose our future selves. In his oration *On the dignity of man*, Pico della Mirandola puts the following in the mouth of the Almighty:

> We have given to three, Adam, no fixed seat, no form of thy very own, no gift peculiarly thine [. . .] A limited nature in other creatures is confined within the laws written down by Us. In conformity with thy free judgment, in whose hands I have placed thee, thou art confined by no bounds; and thou wilt fix limits of nature for thyself. [. . .] Neither heavenly nor earthly, neither mortal nor immortal have We made thee [. . .] though mayest sculpt thyself into whatever shape thou dost prefer.[206]

Human life form is such that, if we consider a plant or animal life form or even Aristotelean nutritive and sensitive souls as a standard, we have no definite form. This is because conceptual mind is able to shape itself, the rational conceptual *bios* is capable not only of using its *zoe* but changing it.[207] We can conceive of possible future selves, consider them, make a decision based on conceptual criteria we hold at the moment, and outline a plan to work toward it; later new decisions can be made and a new course of action adopted. The conception of such future self can be made with different criteria in mind, involving good looks, physical fitness, society to live in, career to pursue, intellectual prowess to acquire, and much more – potentially infinite number of options and combinations. We also conceive of changing society or all of the humankind. Even the physical body is no longer a limitation: technologies expanding the reach of our senses, magnifying our physical power, or dramatically speeding up travel have been with us since pre-historic times; enhancing our physical bodies are newer yet seem promising. Life's self-construction through metabolism in discursive beings acquires a new, radical dimension: being able to re-shape itself based on a preferred concept of the self and following a plan outlined by the living organism. Here the *bios* changes the *zoe*, at times literally: the life form of human biology has the ability to transcend itself, both at an individual and at the species level.[208]

For non-verbal animals, species being is physiological: they uphold it through reproduction, this way ensuring that the proper genetic endowment, together with its instinctual component, is passed to future generations.[209] In some animal species we see some sort of learning, whereby certain associations between internalized images might be fostered among the young, for example, demonstrating hunting practices in big cats and relations between different individuals in primates. However, this learning plays a limited role in animal's life: in most cases, the instinctual endowment will compensate for the lack of exposure to specific behavioral patterns that could have been passed on by the adult individuals. With humans, much of our species being, of what we are, is defined by the behavioral patterns characteristic of the societies in which we live. Out species being is no longer merely our biology but our history, as Hannah Arendt noted.[210]

At the first glance, it might seem that the cultural endowment of human groups takes the place of animal instincts. This, however, is too simplistic of a picture. Instincts guide choice by setting goals and providing some specific behavioral patterns useful to achieve them,[211] and non-verbal animals equipped with the instincts have no choice in the matter of neither goals set by instincts nor particular sequences of actions they need to undertake in order to achieve these goals. A weaver bird is born both with the goal of building a nest and with a specific pattern of tying a blade of grass in a knot – the choice it has is not whether to build a nest or not or whether to tie blades of grass in this or another way, but where to build the nest and what blade of grass to pick. Humans, on the other hand, can consider both goals and ways of achieving them discursively, with none being pre-determined by out physiology. What human culture provides are not behavioral patterns to exercise but mental contents to consider, be those moral norms, traditions of dress, knowledge about the world, methods of inquiry, knowledge of math, or skills at repairing airplanes. These stand before each individual's conceptual mind as something to consider, and it is ultimately up to each individual to decide whether to follow them or not. In many cases such decision is to be made by a human agent aware of the consequences that might be expected to follow, brought about by those who make different decisions. It is also the case that the vast majority of these contents that shape choice alternatives are not developed by the individual alone but are products of societal effort, of human culture. However, the choice is still particular individual's choice. Whatever contents culture provides to us is, differently from genes, subject to the scrutiny by the hosting rational mind.[212]

Human beings can also try and establish new cultural norms, and at times such projects succeed; in fact, many major components of human culture have names of particular individuals attached to them, and philosophy is no exception. It is in this sense that we are cultural animals, beings whose "'essential' self is not one beyond cultural shaping, but one which is culturally shaped in a specific, self-reflexive way."[213]

The effects of our conceptual, enculturing and encultured nature can be conceived of as a *downward pressure* exercised by our free life form on our biology.[214] While mind grows out of life's biology, at its conceptual stage the causal influence is being exercised not merely bottom-up, with biological composition of an organism precluding some and limiting other choices, but also top-down, with minds playing a determinative role toward the very body that gives rise to it. Or, to formulate it differently, a living being acquiring freedom to determine itself more fully.

This downward pressure highlights another important aspect of human freedom – the inevitability of choice. Humans cannot become free of their conceptual nature. At the most basic level this is evident from our inability to ignore conceptual meanings our mind extracts from images. In his famous study that gave rise to a veritable industry of psychological research, Stroop asked participants to name the color of the ink in which color words had been printed: red for a word 'green' printed in red ink, for example. He found that the response time in cases where the ink color did not match the meaning of the color word was consistently slower than in cases where the task was to read the color word.[215] Here abstract, conceptual meaning interfered with a simple perceptual task, despite conscious effort made by the participants to perform it as required. Needless to say, the task would be easy and quick for shapes that are not associated with meaning, e.g., color words in a language the participant does not understand. This experiment, since 1935 repeated in many variations and in virtually every human demographic, demonstrates that we cannot shed our conceptual nature. Our minds extract meaning as automatically as the weaver bird ties a grass blade. We are free to act or not to act upon this meaning, yet it is there to deal with.

We cannot avoid making choices. A human being is, as Sartre famously noted, "condemned to be free."[216] When we choose to ignore certain alternatives, it is still a choice, as we could have done otherwise. The same applies to a refusal to consider a certain situation in depth: it is a decision we make. In this regard, it is interesting to consider the impact of repetition as creating an illusion of truth, also known as the Availability Cascade.[217] As many propaganda men and advertisers know, repeating the same message over and over again gives it more of a convincing force: people tend to believe such messages more than those repeated only a few times or just once. Moons and his colleagues decided to study this effect further, and indeed found that it obtains: repeating an argument causes higher acceptance rates. However, among the participants who paid attention to the argument and found it lacking, repetition had no effect: once they saw it for what it was, no amount of merely voicing it again would make them believe it was a good argument.[218] In other words, when the participants *choose* to pay attention, well-established psychological effect gives way to the power of reason; yet when the choice is made not to exercise it, the outcome is different.

The downward pressure works also in another, enculturated plane. Our conceptual minds are open to any sort of concepts, including those that suggest certain types of "true nature," or way of life we should strive toward or "return to." One might choose to believe, by accepting a cultural idea, that culture is a superficial layer that is not reflective of our true selves (the conception of the true self being cultural as well), and we need to strive toward greater simplicity, abandoning all or most of technological and cultural achievements and listening to our supposed "animal nature." Yet this is impossible: our nature is to be designed by ourselves, "true nature" being no exception. This is one of many possible ways for us to conceive of our self and develop it accordingly. By developing new conceptualization and changing the old ones, human society creates the paradigmatic framework through which we look at the world, ourselves included, a framework which structures the choosing of the individual. The very notions of personhood and life are examples of such concepts. When we conceive of life not as a locus of freedom but as a resource to manage and take care of at the societal level, the type of decisions we will make, the kind of criteria we will use will be different from those that we will employ if thinking of life as first and foremost the seat of subjectivity and will. For example, Illich notes that the "ominous power of modern institutions consists in their ability to create and to name the social reality which the institutions' experts need as the substance they manage."[219] Given such conceptual scheme, decisions regarding the welfare of an individual, a person who is now considered to be an instance of life, can be made without much heed to the opinions of the individual.

One perennial philosophical problem also might be a byproduct of our reflexive rational nature – the mind-body problem. Due to our conceptual reflexivity, the human organism is capable of looking at aspects of itself, its body and its thinking processes, as objects. This way, the organism uses mind to conceive of its body, and sees it as an object that opposes its I, the subject, as something separate from it. If not for the ability to look in a similar way at the mind, the mind-body problem would never be born, as it requires to conceive of mind as well. However, the reflexivity does not stop with a body, and thus a conception of an immaterial or peculiarly material, e.g., similar to smoke or air, mind is born. Then, the mind-body problem arises: how can such a mind be connected to a body? All this, thanks to the reflexivity or organisms endowed with discursive intelligence.[220]

Another outcome of misusing rationality, as unfortunate as it is common, is assigning undue significance to symbols and signs. Associative minds incapable of conceptualizing can only link images to other images and then respond to one as if exposed to another, or, to be more precise, anticipating the exposure to another. With human, we conceptualize the connection. When we link an image of an animal to certain character traits we wish to possess: lions and bravery, ravens and majesty, etc. – we understand full well that there is no "natural" tie between the two, that it is a creation of our mind. We create flags to represent, i.e., invoke thoughts about, countries or political movements, where there is a sensory link between, say, red and fight for independence as both are tied to bloodshed – yet we clearly understand the symbolic, arbitrary nature of choosing this and not another color, e.g., not brown for the clay on which the beloved Motherland stands. Certain verbal signs, e.g., a name, is also being linked to concepts, e.g., those of persons or places. Similarly, discursive mind enables us to connect between events by developing concepts of magical relation or relation of representation. The *post hoc, ergo propter hoc* fallacy consists not in simple conditioning but in conceptualization of the observed temporal sequences, e.g., through mythological explanation. Yet then, once the mythological explanation is already conceived as such, an odd thing happens: we start assigning to the myth itself, as we do with some symbols and signs, conceptual significance tied to the respect and dignity of what they represent – the latter being concepts as well, of course. This way, any harm caused to the symbol or a sign is conceived as a harm to the group or an individual that subscribes to this relation. Many a conflict ensue, as Lorenz bitterly noted.[221]

Perhaps even more dangerously, people can suppress empathy in order to comply with their chosen self. Chimpanzees, known to feel another's pain, demonstrate consolation behavior, where they show compassion toward a suffering other, e.g., an individual defeated in a fight.[222] Since chimpanzees lack language and with it conceptual thinking, there is no reason to believe that they arrive at the need to show compassion through deliberation, formulating behavioral options and weighing their merits in accordance with abstract criteria.[223] As humans have similar neural equipment that underlies empathy, it makes sense to see compassion and associated behaviors as having their roots in urges passed down from the primate ancestry we share with the chimps. Yet, using our conceptual minds, we can suppress such behaviors no less effectively than an urge to yawn in formal settings: we can get ourselves convinced that compassion is inappropriate or damaging in certain situations, and demonstrate total lack thereof, if not considerable cruelty.[224] Milgram's experiments, where participants administered what they thought were extremely painful electric shocks, even though they were free to refuse doing so at any point,[225] are rarely considered in the context of freedom. Yet they provide, perhaps, the clearest example of how far conceptual human choice can go in suppressing the call of our neural endowment. Of course, a quick glance at any history textbook would yield even more striking examples. Interestingly, similar experiments were performed in rats in the late 1950s: the rodent subjects had to make a choice between pressing a lever to get food and cause another rat to suffer electric shocks, or refrain from causing pain to another rat and stay hungry. While some rats ignored the suffering of the other, many stayed hungry for as much as two days and even then pressed the lever at a reduced rate.[226]

This leads to another important aspect of conceptual human freedom: its ability to suppress its own expression. Humans can develop new concepts, including those that deny the reality of freedom. For example, people might postulate the illusory nature of free choice, seeing humans as driven by instinctual urges or being merely tools at the hands of gods. We can choose to believe in such concepts, and this way, paradoxically, to cease believing that we actually can choose.[227] Such beliefs, while not eliminating out ability to choose, might limit the practical range of choice: choosing not to foster the range of available alternatives will lead to situations where fewer courses of action stand before the deciding mind. Somebody who decided not to pay serious attention at science classes would have no possibility to pursue a career in nuclear physics later in life – not because it is principally impossible but because this alternative would not be available to him in the circumstances of the society he lives in. Same applies to cases where individuals are not exposed to certain intellectual and cultural contents through no fault of their own, e.g., denied quality schooling by the group to which they belong, be it society or family.[228] This is similar to Dennett's design space,[229] but at the micro level of individual freedom: every choice has a potential to impact the range of future choices, and limitations imposed by the environment restrict not only the number of options available at this moment but also further development of individual's conceptual structure and thus the choices this individual will be making in the future.

How does this square with the notion of absolute freedom, at least a grain of it, that is coveted both by the libertarians and by the common view of freedom? Such freedom is impossible on the account presented here: living agents are ever constrained in their choice. It is, however, meaningful in an important way: as an ideal. Absolute freedom, as a concept discursively intelligent animals can comprehend and hold as their developmental goal, impacts behavior: having this ideal, human beings can try and measure their practice against it, strive to maximize their freedom even though knowing full-well that absolute, unlimited freedom is impossible. We can restructure political and economic practices, social institutions and family, education and medical care to enable meaningfully freer choice. With the downward pressure described earlier, we can achieve greater freedom, eliminating more and more barriers, equipping people with more and more tools of choice and presenting them with more available alternatives. The ideal of absolute freedom becomes then an asymptote toward which the curve of human development can strive to approach: never reachable yet closely approximated.

This is, then, the playground of human freedom: potentially infinite space of conceptual consideration, in practice limited by prior decision history, physiological confines we are yet to overcome, and the limitations on our choices imposed by the societies we live in and by other individuals.

ANIMAL CHOICE VS HUMAN FREEDOM: A COMPARISON

Mind appears to be the central aspect of the animated, as opposed to vegetative, life form: supported by animal's nervous system, it pervades all of animal's organism while not being associated with any specific organ exclusively. Mind is what makes self-determined choice possible: to choose, one needs to represent the alternatives, make a decision, and carry it out. This requires central control over the organism that is capable of representation, preference, and activating the organism to act on the preference. Thus, mind is the main way of use animal's *zoe* makes of itself, and the main form of choice. Because of this, the form of mind, the way it operates will define the character of choice.

Animal choice is exercised over intuited contents without mind's conceptual awareness of making a choice, i.e., of being engaged in a process of selecting one behavioral alternative out of several available: such awareness requires conceptual reflexivity which is lacking in non-verbal animals. The same deficiency precludes non-verbal animals from being able to have complex mental structures where cognitive contents relate to each other based on aspects of meaning that do not have to be rooted in sensory properties of mental images. Instead, non-verbal animals are limited to mere association of more or less generalized representations. These two factors make non-verbal animal's choice limited to one dimension – a spatio-temporal dimension of associated intuited content. Thus, animals have an environment in which they act, environment populated by mental representations of sensory images. It can be quite rich and include even rudimentary theories of mind,[230] yet still limited to this one dimension of choice and to the environment the animal agent encounters.

Humans, on the other hand, have a world – not only a natural environment that can be intuited, but also the one that we conceptualize, the space of meanings we assign to intuitions we have and the links between them. Thanks to the hierarchical and abstract nature of concepts, human world is much richer structurally and content-wise than animal environment. This world is also one of our own making: we keep changing and expanding it, altering existing meanings and producing new conceptualizations. Thus, human freedom is not only multi-dimensional, but the number of dimensions keep changing.

The type of alternatives non-verbal animals and humans have to choose from is also different. Animals choose between future states of affairs distinguished by their intuited features. The main driver here seems to be animal's desire for this or that sensory outcome: feeling pleasure at getting food or receiving attention, for example, or alleviating the urge to console another member of the clan. Humans can too, of course, limit their choosing strategy to similar consideration, but this in itself would be a choice, as other routes are open before us. When we choose between different courses of action, we can compare them on different scales: utility, morality, familiarity, complexity, veracity, risk, etc. We also can give preference to this or that criterion. It is up to the deciding human being which scale to use, even though the criteria themselves are conceptual contents in our minds, and the acquisition of these contents frequently depends on the culture and society we live in, the position we occupy, and the life experience we have.

Along similar lines, types of concerns that shape choice are different for humans and animals. Thanks to their associative minds, animals can *expect* certain outcomes to come to pass. Together with desire, this can give rise to strong emotional displays and mobilize animal organisms to act. Humans, on the other hand, can *hope*:[231] express to themselves the desire for certain outcomes with which they are emotionally engaged, evaluate such outcomes as possible yet not highly probable, and thus develop certain cognitive disposition that is distinct from expectation or belief.[232] It is the presence of this evaluation of probability, as opposed to mere desirability, of the outcome and forming an attitude based on the results of such evaluation that makes hope distinct from expectation and as impossible for a non-verbal being to demonstrate as belief. To relate to the desirability of a future event, a mind-endowed organism needs no more than an image of this event. After learning to associate the coming of a laboratory associate with food, Pavlov's dogs were perfectly able to associate the events temporally related to it, e.g., the ringing of a bell, with the pleasure of eating. However, to evaluate the probability of a future event, the agent needs to have a concept of likelihood related to the concept of time, and then relate to that the desirability of the event. When the event is in the future, the desirability is high, yet the probability is low, we would hope rather than, say, expect.[233] Similar analysis can be applied to other concerns: love, duty, commitment, conviction, faith, and more. All these impact human choice yet do not figure as factors in animal choice.

Because of the much wider scope and deeper reach of human freedom, humans are also able to consciously change their own world and themselves in ways detrimental to the exercise of freedom. Non-verbal animals' choices can, unbeknownst to them and with no conscious intention on their side, change the course of their life and, in some cases, the lives of their descendants: wandering into somebody's backyard for a stray cat might well change the course of a cat's life. Joining human nomads had profound consequences for the *Felis catus*. The choices humans make, besides having a much wider range of consequences, known as unknown, might consciously limit human freedom, the ability to make choices, as discussed earlier. Moreover, certain human choices can lead to extinguishing most, if not all life on Earth – a feat that is inaccessible to any other animal.

Yet despite all the differences, animal choice and human freedom share a number of important aspects. The most important, indeed crucial of these is the nature of freedom as self-determined choice between alternatives. Considerations of scope aside, this aspect is common to man and beast. As I will argue in Part II, the presence of choice in animals is a pre-requisite of freedom in humans.

Another aspect is also common in non-verbal animals and discursive humans: our choices impact our future ability to make choices. Non-verbal animals are not aware of the possibility of such impact, as it requires conceptualization of time, of choice, and the relation of possible influence. Linking the image of biting another animal to the image of a repercussion that has occurred in the past requires no more than the ability to associate mental contents. Yet grasping the conditional relationship of "if I attack this animal now, in the future I will have no choice but expect a reaction similar to what I am going to experience now, and thus the repertoire of my choice alternatives will be limited" requires understanding the relation between one choice to another, and specifically present choice to future choices that are yet to be made. If situations of choice register merely as images, mental imprints of the feelings associated with choices, such analysis is impossible. To relate choice as a situation of selecting between alternatives in relation to the arrow of time that has no sensory content, and to see this relation in the aspect of present choice influencing future choice, the agent needs to conceptualize choice, time, and influence. Humans are sometimes aware of the impact of past and present choices on future choices and sometimes are not aware of it, as we quite often cannot predict the influence of the consequences of our actions on the possible scope of our future choices. Sometimes we simply do not care to think about it. But we are capable of understanding this relation and planning accordingly, as we often do, but perhaps, not often enough.

This lead to another commonality: in humans and in other animals the history of individual choice accounts for much of individual differences. A pet dog who chose at a certain point to try and fight it out with a feral cat might avoid even a whiff of feral cat's smell in the future – something that a dog who has never experienced such drama might not do. A human being who decided to study philosophy yet skip most of the readings would be quite different from his classmate who chose to read them all – for example, in terms of the number and quality of decision paradigms at her disposal whenever questions of life come up, not to mention passing exams.

Neither discursive humans nor higher animals, those capable of choice, can avoid choice. A bird that needs a twig to complete its nest has to choose a twig. A human who is considering to buy a book also faces a choice, whether to do it, what book to buy, where to buy it, etc. While the range of bird's choices is more limited, both have to choose. The emergence of choice and the development of life are bound together, with the evolution of *zoe* leading to the unfolding of greater freedom, and the *bios* of self-determined choice impacting *zoe*, most strongly at its human stage.

As beings that cannot avoid choice, both animals and humans might face situations where the tools they have at their disposal to assist with choice would not help. A non-verbal animal might encounter a sensation it cannot ignore yet fails to associate with anything else in its store of memories. This sensation might also fail to trigger any instinctual behavioral pattern. In this case, the animal is likely respond with fear – perhaps, the most evolutionarily sound response in this case.[234] A human being who encounters something he or she cannot conceptualize, put in its existing conceptual scheme or extend it consistently with what it already has to accommodate the new experience, might well have similar response: Arjuna's reaction at seeing

Krishna's true form, as described in *Bhagavad Gita*, suggests just that in a vivid form.[235]

Freedom is a continuum. Its seed exists in the plant life, yet only *in posse*. Animals, when they acquire the necessary cognitive prerequisites, start exercising choice – very limited at first and then expanding with the increasing sophistication of their ability to store, generalize, associate, and retrieve images. With the development of language, human freedom with its virtually unlimited scope is born. Similar process takes place at the ontogenetic level. In non-verbal animals capable of choice, it develops along with their sensory and cognitive systems, particularly the memory. In humans, pre-linguistic children are no more capable of choice than non-verbal animals, and only with the mastering of language do they become free in the human sense. Yet this freedom becomes richer, more capable with the development of the system of knowledge, sensory or conceptual, each individual possesses. Here non-verbal animals and discursively rational human beings are similar as well.

Part II

THE EVOLUTION OF FREEDOM

Chapter 2 establishes that animal life makes it possible for freedom to exist in a determined physical world. First, non-verbal animals endowed with representing consciousness[1] exercise choice between alternatives. Then, humans come to have a wider and deeper form of freedom thanks to being endowed with discursive metacognition and thus being capable of envisioning a much richer scope of options to choose from and a variety of ways to make choices.[2] The purpose of the second part of the book is to provide an evolutionary account of the development of freedom: to demonstrate how life, having an inherent potential to evolve, gives rise to different forms of self, develops the components necessary for free choice, and then leads to discursive intelligence that culminates in the human form of freedom.

In what follows, I will argue for the directionality of evolution toward greater freedom and for the mutual influence between an organism's biology and its capacity for choice, where the influences of choice will have a serious impact upon the character of evolution in non-verbal animals capable of choice and change the character of evolution in humans. Alongside this development, the influence of genetic endowment on the organism's way of life will be diminishing, and the scope of choice will be increasing.

Chapter 3

The Philosophy of Evolution

THE TASKS BEFORE THE PHILOSOPHICAL EXPLORATION OF THE EVOLUTION OF FREEDOM

Philosophy addresses fundamental questions of knowledge and being. When it asks these questions about itself, it attempts to operate without an externally given content or basis. When it asks such questions about other domains of being and knowledge, they can be described as meta questions, questions about other questions and about potential answers to these questions – their coherence, validity, scope, etc. For each area of human knowledge, existing as well as possible, such questions can be posited. The domains of inquiry attended by empirical sciences are no exception. For example, in the case of physics philosophy will ask questions regarding the nature of physical phenomena, the possible validity of the laws of physics, the existence of time and space and their nature, and thus define the scope and the terms on which empirical exploration of the material world can proceed, as well as the frameworks in which its findings can be interpreted. In biology, philosophical reasoning "can explore what is constitutive of life in general as well as what are the fundamental particular forms that life can take wherever in the universe living things may arise," no matter what the particular biochemical instantiation of life might be.[1] This way, we can think of universal biology that would address processes essential for life.

In the case of evolution, philosophical reasoning is tasked with not only exploring what forms life can take, but what are the developmental paths the evolution of life can follow: what are the necessary pre-requisites for the development of certain life forms, what are the constraints on such development, and whether it has any teleology. This way, philosophical exploration of evolution is a department of universal biology that addresses the questions about possible sources of life and its development.

When exploring the evolution of freedom, philosophy needs to find out how the features necessary for the development of animal choice and then human freedom can unfold with the progression of evolution, as well as address whether and how biological evolution and choice can influence each other: the nature of such influence and the forms it might take. Then, philosophy should consider the empirical findings as they pertain to the evolution of freedom and interpret them in light of the conceptual framework it has developed, as a test for the ability of the philosophical evolutionary theory to provide a paradigm for understanding biological evolution.

PHILOSOPHICAL UNDERSTANDING OF THE EVOLUTION OF FREEDOM: THE SUGGESTED COURSE

Dobzhansky noted that "[m]an's humanity and his animality are not independent or kept in isolated compartments; they are interdependent and connected by reciprocal feedback relations."[2] This is something that is hard to miss if one thinks about the human life form. Whichever way we define 'man's humanity,' it will influence our biological nature: our thoughts have great effect on our emotions; our science impacts our health, and so does our way of conducting business; our culture impacts mating preferences, and thus the genetic endowment we pass to the next generation; and much more, all the way to altering the biochemical processes that sustain us and ending our lives for the sake of certain ideas. On the other hand, it is clear that our bodily needs impact our economic lives, our hormones impact our emotions, the shape of our body and its way of functioning has a lot to do with the laws and institutions we put in place, the features of our neural systems play a great role in how we communicate, etc.; no one would doubt the influence of our biology on the way we live.

This principle is not confined to the human species; rather, it seems universal for all living beings. The way of life of a living entity, or the use it makes of its body, always impacts its biology, both of the living individual and of the individuals that will evolve using its genetic code. When we consider plant life, the notion of the way of life is merely formal: plants lack choice, and the hand of fate decides what is the environment that will impact plant's biological life story: would it grow tall to get its share of the sun if its seed chances into a thick grove, or spread its branches around if it grows in an open field; would it pass its genes further if it reproduces sexually and there is a plant of a complementary sex in the vicinity or not; and more. On the other hand, once choice becomes real, with animals who are endowed with representing consciousness,[3] the choices the animal makes will impact its life and its progeny: the food that it eats will influence the development of animal individual's organism, the places it travels will lead to encountering experiences that are likely to shape its future behavior by forming associations that will be remembered, and the mates it seeks will play a crucial role in the genetic composition of its offspring. This, however, is only part of the interplay of the life form type and species' biology. Animals capable of forming groups and communicating with group members through signaling[4] will be able to transmit certain habitual behaviors to later generations, develop behavioral patterns that are sometimes misleadingly referred to as animal culture.[5] Evolutionarily, the capacity to transmit behavioral patterns to future generations is important for survival in these animals, and thus it will be favored by natural selection, perhaps blind with regard to any purpose but very sensitive to its subjects' way of life.

Once the ability to choose between alternatives makes its appearance, it becomes a factor in natural selection. Animals can now consider, however unsophisticatedly in terms of the criteria and comparisons involved, different options and select those that seem to them to be aligned the most with their interest, defined here simply as pleasure: satisfaction of hunger and thirst, lust and emotional need, etc. This ability can be advantageous for survival, as I will try to argue later, or disadvantageous; or irrelevant – all depending on particular circumstances. Yet this is something that plays a role in natural selection. Moreover, the choices animals make become a factor too, as they change the options available for further evolution, or alter the confines of Dennett's design space.[6] Different groups of animals of the same species might develop different ways of coping with challenges, e.g., with getting food. For example, one group of primates might start using sticks for fishing for termites, while another, with the same cognitive and physical endowment – to try and destroy anthills. Over generations, given that both groups survive, multiply, and pass their habits on, natural selection in the second group is likely to favor strength, while in the first one – fine motoric skill. This, in turn, might lead to uneven evolutionary development of the neural and motoric components responsible for different types of skills – which will lead to new capacities, pulling the groups further apart socially and genetically.[7]

Similar selective processes would apply to animal learning. The ability to retain images of stimuli, link them to other retained images, and retrieve them upon encountering sensations that give rise to similar images enables conditioning, operant as classical. However, for an animal capable of choice learning is not limited to conditioning. In many *in vivo* cases, conditioning is not as neatly arrayed as in the lab. Usually an animal would encounter conflicting urges, e.g., an instinctual one and one that is triggered by a conditioned response, or two conditioned responses that are tied to stimuli being encountered – only in Pavlov's lab is there just one stimulus available while the rest are being held in abeyance. In this situation, the animal needs to choose.

An argument can be made that it is the strength of the urge that makes the animal choose this or that response. However, it is impossible to define what does the strength of the urge mean here. How would the memory of pain encountered last time the sensation that is being felt now was present compare to the empathy or lust that are being felt at the same time? Would an animal behave in the same way in all sufficiently similar situations? What does it mean to be sufficiently similar when sensations are concerned? None of these questions has a definite answer wherever we consider the natural environment rather than a sterile *in vitro* setting. Imputing to animals behavioristic mental mechanism lacks not only empirical support,[8] it is more of an axiom we are trying to impose on animal behavior than a principle that is supported by argument and observations.

Non-verbal animals do not have access to universal normative standards – they cannot choose this or that option because they think it to be more merciful or just. They can relate to alternatives based on the imagined outcomes that are more or less pleasurable, more or less satisfying in terms of urges, emotions, etc., sensory parameters. Yet we have no reason to see the amount of pleasure as the basis of animal choice, as has been argued in the preceding paragraph – it is the context in which their choices are made. We cannot discern a specific measurable criterion which the animal would necessarily follow, grade the options along a corresponding measurable scale, and argue that this gradation determines what the animal will do. Similarly, in the case of discursive human agents, we can outline the context of choice, which, in addition to the animal's pleasure and pain, emotions and urges, also has the domain of universal normative standards, where the latter provide a framework through which the former are considered. Yet we cannot discern what are the specific decision criteria this or that human being would follow – the freedom to choose is theirs.

This way, the capacity for choice transforms the character of evolution in two ways. Firstly, it makes animals capable of choice active agents of evolution rather than its mere patients: now their choices matter, and by perceiving the consequences of their choices and acting upon them they change the evolutionary course of their progeny. The animals themselves are not aware of this relation between their choices and evolution of their species, as such awareness requires abstracting from images, the aboutness of linguistic intelligence that is necessary for considering relations between events as an object of thinking. Secondly and perhaps crucially for the direction of evolution, starting with the very evolutionary moment when choice makes it phylogenetic appearance and proves useful for survival, the neural structures that support it become advantageous for the bearers of the genotype that encodes them. This means that those individuals who are better at considering alternatives, relating them to their own good linked to survival and successful reproduction, and choosing the best option will survive and reproduce better than the rest. A bird that can generalize better over the images of blades of grass would be capable of picking a blade that is different from the one type that is instinctually encoded; therefore, it will be able to build a nest when the usual kind is unavailable, and thus survive. Better yet, a bird that can generalize over the images of nests and thus be capable of choosing ready-made shelters fit for laying its eggs would have a better chance, especially if it stumbles upon a battlefield strewn with dead soldiers' helmets the wearers of which had managed to kill each other thanks to their nations' lofty ideas made possible by linguistic intelligence.

The neural structures that underly the consideration of alternatives and thus choice can be understood as engaged in data processing, where data is seen as generalities abstracted from images. The softness and size of a grass blade, the shape of a concave object and its hardness – all these can be seen as aspects of images. Processing them means relating these pieces of data to something else, in the case of choice according to the animal's own good defined as pleasure – to such good, or interest. At this stage, no consideration of the relationship between data pieces themselves and the good that is used as a criterion is possible – there is no metacognitive aboutness to mind yet. However, if natural selection starts favoring these structures because of their contribution to survival, their processing capacity will increase with phylogenetic development for the same reasons anteater's tongue probably got longer with anteater's evolution. Among other things, mind's ability to range over its contents will be improved: not merely a blade of grass but any similarly structures object, not a set of twenty signals but a set of two hundred signals, etc.

This increase in processing capacity leads to a point where not only generalized aspects of images but the very process of generalization itself will become an object for mind's operations. The more images are generalized, the larger is the distance between their physical aspects and their meaning to the animal, the higher is the probability that at one point the relationship of generalization itself will become something to consider.[9] Perhaps, something like that happens sporadically here and there with individual animals that have advanced cognition. When this becomes available to enough members of an animal group that practices signaling and thus operates with symbols, they might start using their existing abilities differently, and begin the transition from symbols to signs and ultimately develop language.

Language opens a new and ever expanding horizon of human freedom.[10] The exercise of this freedom changes the nature of evolution. With language, humans acquire the ability to assign names to any mental content, sensory as well as conceptual, and relate them to each other, including making judgements in light of various criteria, by using Chomsky's Merge operation which becomes, in this sense, a vehicle for metacognition.[11] This makes possible for humans to develop normative thinking: being able to operate with imageless mental contents is a necessary condition for comprehending universal norms the mental representation of which is non-sensory. This way, humans can also argue about norms, thus giving rise to differences in opinion and different cultural practices. They can now consider not only doing things, choosing between alternatives that can be envisioned, but ways of doing things, thus creating new alternatives deliberately. These ways become an important factor in the structure of evolutionary pressures, at a certain point – more important than natural conditions that affect food supply, availability of shelter, and climate for non-verbal animals. In this sense, cultural selection can be seen as competing with the natural selection at the beginning of human history and gradually displacing it as the major factor impacting survival; with the development of technology the rate of this displacement increases: it makes cultural norms more effective in impacting evolution. For example, with the advance of medicine, not passing your genes and cultural heritage on becomes less of a matter of physical fitness and more the question of marriage preferences, success in society, and sheer luck of not having your country turn into a war zone. The impact of culture on evolution can proceed at first without explicitly considering evolution, yet humans can also think about evolution: try and understand its workings and then attempt to change its course as they see fit.

All this points to a certain directedness of the evolutionary process. With the advent of life, the first seed of freedom is sown – the separate self appears. Given the "descent with modification," to use Darwin's term,[12] the pressure to survive will lead at a certain stage to the appearance of motile, perceiving, self-controlling animals.[13] This move seems to be following from the very nature of life that leads to evolution. Animals will be subject to evolutionary pressures to utilize their perception, motility, and self-control better in order to survive till the reproductive age. For animated organisms the way to do it is to develop choice – which is fueled by the data processing capacity and thus favors genotypes that ensure better data processing. Thus, the very evolutionary nature of life will cause an increase in organism's data processing capacity which at certain point will give rise to metacognition and to language, the enabler of human freedom. I will argue later that the increase in the ability to process sensory data, store it, and manipulate it contributes to survival, and thus is favored by evolution. This increase will lead to this process turning upon its own workings, becoming reflexive – the development of discursive metacognition, the language-enabled ability to go meta on any mental content, including the one that has no sensory component. This way, we can see the evolution of life as the unfolding of freedom, as Hegel's view of mind can be interpreted.[14]

In what follows I will try to substantiate the claims outlined here, after addressing the pertinent questions of biological evolution.

Chapter 4

Biological Evolution

LIFE AND EVOLUTION

Evolution is a historical process of development, where various characteristics of the entities that are subjects to it change over time. As such, the term has been widely applied in various spheres, including to then whole of nature under the heading of cosmic evolution.[1] This wide application, however, makes the term too broad to distinguish between different types of development over time, as it becomes almost synonymous with change. The scope of evolution relevant to freedom is evolution of living entities, or biological evolution.[2] Consideration of change in this case yields a number of characteristics that figure in the most common definitions of evolution.[3] Firstly, biological evolution is interested in the changes that occur in the transition from one generation to the next, not the changes the same organism or multiple organisms demonstrate over their lifetimes. Secondly, evolution addresses changes that are characteristic of a large number of individual organisms: one or two larger beaks in the generation of a hundred thousand finches or a better ability to communicate in a few primates out of many hundreds of thousands would not count as an evolutionary change. Thirdly, the new characteristic should be passed to the next generation in such a way that it becomes a feature of a large number of individuals of that generation; I am deliberately not specifying the way this change is passed on – I will argue later that in the course of the biological evolution new methods of passing newly acquired characteristics to the next generation appear. As a result of the type of evolution characterized by these three aspects, different species are formed – new groups of organisms that are distinct from other groups in terms of bodily composition, demonstrate different typical behaviors, and are reproductionally isolated form other species.[4]

Theodosius Dobzhansky, one of the authors of the Modern Synthesis, a conceptual framework that reconciled Darwinian evolution and genetics, famously argued that "nothing makes sense in biology except in the light of evolution, *sub specie evolutionis*."[5] The reason for that is that the unity, or the commonalities between very different organisms, e.g., genetic similarities between humans and broccoli, is comprehendible only if we assume common ancestry; and the diversity makes sense only as an outcome of adaptation to different environments. One can argue, however, that this view uses evolution as merely an intellectual crutch that helps us to understand the world as it happens to appear to us. As such, the evolutionary paradigm can be as restricting as it is enlightening: since it helps us to understand the world, we might get into the habit of ignoring anything that it cannot explain, use it as a guide for discovery – and end up looking for a lost item under the lamppost not because it was lost there but because this is the only place lit. Thus, if we are to stick to evolution, we ought to establish it on a basis that would be firmer than intellectual convenience.

Fortunately, such basis exists: life cannot but evolve.[6] Life's self-constitution relies on its organic unity, metabolism, and reproduction.[7] As unified organically, a living entity is different from its environment in terms of the combination of processes that sustain its functioning. As metabolizing, it depends on its environment for the inputs it needs for its growth and sustenance, e.g., nutrients. Thus, each unified autopoietic organism metabolizes on its own terms: the way it interacts with the environment is defined by the biochemical processes that constitute it. If two organisms are different from each other in terms of such processes, and the difference is relevant to the way they obtain the necessary materials and use them, their survival and reproduction chances will not be the same: if of the two amoebas one is quicker at absorbing the molecules it needs to survive and then divide, given the scarcity of such molecules it is more likely to survive; same considerations applies to all organisms, uni- or multi-cellular. Reproduction provides both for passing on the features that assist in survival and for making the organisms of the next generation different from each other,[8] almost always – very slightly different. Darwin's descent with modification which can be rooted in passing on genes that define how the organism's physiology will turn out, factors affecting the expression of genes, and social learning in more cognitively sophisticated organisms all instantiate this principle in life on Earth. Reproduction ensures that each autopoietic organism is unique, makes it a hypothesis offered a chance to be tested in the court of survival and give rise to the next generation that is better equipped to withstand the tests to which their parents were subjected thanks to the beneficial traits they inherited. Thus, reproducing autopoietic life is bound to evolve.

THE PRINCIPLES OF EVOLUTION

The idea of evolution as the accumulation and inheritance of change in living organisms over time is quite old. Empedocles suggested that living organisms were formed as a result of recombination of parts, and out of all kinds of monsters those who were fitted to survive indeed survived and gave birth to offspring.[9] Aristotle was aware of and argued against the accounts that suggested the decisive role of change in generating varieties in the shape of body organs and being fit for use as the decisive factor in what survives.[10] Darwin's own grandfather suggested evolutionary change in animals.[11] There were other philosophers and scientists who proposed evolutionary ideas,[12] yet it was Lamarck who offered the first argued theory of evolution that addressed issues relevant to the consideration of freedom,[13] and Darwin who developed a theory that has been supported, with significant modifications, by empirical research.[14]

Darwin's account of evolution, the one that has laid the foundations of modern evolutionary research, can be summarized in three core principles operating in nature that he discerns from observation: variation, inheritance, and selection.[15]

Variation

The principle of variation states that individuals of each generation belonging to a species differ in terms of morphology, physiology, and behavior. Without variations that can be passed to the next generations, in this or that way, there is no development. Variation is also necessary for the survival of life in a changing environment. For a fixed type of organism, changes in its environment, e.g., temperature, humidity, and availability of nutrients, might spell doom if it is not adjusted to the new conditions enough to survive and reproduce. Having varieties, though, increases the chances that some individuals would be able to survive in the changed environment.[16] Thus, variation is a necessary component of evolution.

One can think about four sources of variation: the set of instructions for organismic development, or genetic code, passed from the parents; the way these instructions are followed, or gene expression; individual differences resulting from the environmental impact; and choices made by the individuals.

The origin of much of variation, most variation in plants and simpler animals, is in the particular structure that supports inheritance in all life of Earth: the DNA, a macromolecule that consists of repeated elements, or nucleotide bases (cytosine, guanine, adenine, and thymine). The arrangement of these bases dictates the material composition of the organism and its chemical processes, and thus underlies it development, growth, reproduction, various other physiological functions, and decay. During the process of reproduction, whether sexual or asexual, some segments of the genetic code might change – a mutation may occur, thus potentially giving rise to different physiological features in the organism to be born. However, mutations are not the main source of variation: most of the genetic code is not functional, not utilized in regulating any organismic processes, so mutations in it would be non-consequential.[17] Since mutations are random, chances are that if a gene segment that is important for organic functioning mutates, the organism will not live. Moreover, important functions are encoded by more than one DNA segment, so if one has mutated, others will support the same function.[18] The main source of genetic variation is reproduction, where the DNA can be altered in two ways. In sexual reproduction the genes from both parents mix, then some genetic segments can recombine again through a process called "crossover."[19] As a result, a unique genetic code is formed, giving rise to an organism slightly different from its parents.

In the 1930s genetics was reconciled with the Darwinian evolution through what has been called the Neo-Darwinian Evolutionary Synthesis, or Modern Synthesis.[20] According to the modern synthesis, the changes in the genetic code accumulate slowly, and thus the processes of change and speciation, divergence of the genetic code to the degree that it gives rise to reproductively isolated, phenotypically and behaviorally different groups, or species, would be very slow as well. This conclusion, however, was challenged in the 1950s by Robert Goldschmidt who suggested that the process of speciation is discontinuous, that changes in key genes that control serious major features give rise to new species through what he called "hopeful monsters," individuals with characteristics that would make them much more suitable for life in general or for a different way of life than other members of their species. Ridiculed at first, these ideas were vindicated by the research toward the end of the twentieth century, when such regulatory genes, or controlling genes, were discovered.[21] Additional research found that speciation can proceed at a much more rapid pace than Darwin himself and the neo-Darwinians suggested: dozens of speciation events have been documented in the last fifty years for multiple animal species: finches, sparrows, salmon, sticklebacks; speciation in microorganisms is even more rapid.[22] The new field in evolutionary biology, evolutionary development, known by a nickname evo-devo, is studying how regulatory genes can cause major differences and re-writing the textbooks created by neo-Darwinians.

The second factor in variation is gene expression. It is not enough for a genetic sequence to be present in an organism to shape its structures and functions, it also needs to be active to produce amino acids, the building blocks of proteins, the building blocks of the body. Almost all cells in the organism include the same genetic code, but thanks to different expression some genes are active in cells that make the materials necessary for creating oak's leaves, and other – its roots. Gene expression might be regulated by environmental factors, e.g., the availability of chemical compounds that would be necessary for the gene to be expressed. This is why the lack of proper nutrients, for example, might cause animals to grow less hair or to have brittle bones: the compounds necessary to have the genetic code creating the proper amino acids would not be there. In other cases, specifically in the organisms comprised of cells with a nucleus, or eukaryotes, certain biochemical compounds might prevent gene expression at the epigenetic level, when the DNA is ripe for transcription, by tightly packaging certain segments of the genetic code and thus not allowing them to use the materials inside the cell to create amino acids. Epigenetic changes happen during the individual organism's lifetime. However, research shows that they can be heritable, even though they do not change the composition of the DNA but its structure.[23]

The third form of variation is an outcome of external conditions that impact the organism during its lifetime. Depending on the availability of nutrients, water, and sunlight plants might be different: larger or smaller, with more or fewer branches, paler or brighter leaves, etc. Similarly, individual animals will develop different phenotypical characteristics based on the food available to them. These differences had been long discounted by evolutionary scientists. With the advent of epigenetics they have taken a new hold, yet their influence is still relatively minor: only in certain cases can environmental conditions cause an epigenetic change in an organism be inherited by its offspring. Still, these changes can be part of the behavioral variation that might well be passed on, though not through the genetic code.

The more sophisticated is the animal organism in terms of its mental apparatus, the more it is subject to behavioral modification by environmental changes. In animals who can retain sensory images and link between them and the intuited ones, exposure to new foods makes animals capable of choosing which food or to avoid. Harsh environmental conditions might make it migrate, for good or only during certain seasons. Environmental factors may change social behavior, make animals fight or stick together. All those variations are subject to the pressures of natural selection no less than phenotypical characteristics.[24] Moreover, many can be passed down to future generations through social learning, as it will be discussed later.

In animals capable of choice, interpersonal variations can be much wider than in those who are not; and the potential scope of interpersonal variation seems to be proportional to the scope of choice. In animals driven by instincts, the scope of choice is relatively minor: they cannot switch the instinct on and off on their own will, yet they can choose, for example, different nesting sites and different materials for nest building. However, in animals who can make wider choices, for example, animals whose life is less regulated by instincts, behavior can become much more diverse as a result of choices made, thus causing a wider repertoire of behaviors and abilities to be subjected to natural selection. A new evolutionary factor is born here – animal choice.[25]

With the advent of linguistic intelligence, the scope of choice explodes and its nature changes, as has been discussed earlier. The amount of interpersonal variation becomes much more evolutionarily significant. The character of evolution changes as well: now humans can control many factors that impact evolution, for example, natural selection in regard to individual's physical health.[26]

Heredity

The principle of heredity pertains to passing the variations characteristic of parents to the offspring. In the sense relevant for evolution, heredity is passing the resources needed for ontogenetic development from one generation to another, resources that ensure the reconstruction of biological patterns characteristic of one generation in the next.[27]

In the case of life on our planet, heredity is frequently identified with the physiological structure of inheritance that relies on reproduction that involves passing parents' genes to the offspring. This way, all variation in parents that is genetically determined will be passed to future generations, even though its expression will be subject to the dominance and recessivity of genes.[28] An important point to note here is that only those characteristics that are supported by the genes will be passed to the next generation by this form of inheritance. Yet what about characteristics that are not reflected in the genetic code, e.g., those that are acquired by the organism throughout its lifetime?

The main difference between the Darwinian and the Lamarckian evolutionary theories consists in their treatment of acquired characteristics. Both Lamarck and Darwin held that features acquired during animals' lifetime can be passed to their descendants; however, the inheritance of acquired characteristics is associated with Lamarck's name,[29] as those following Darwin's path quickly abandoned this approach because empirical data they were familiar with, both observation and controlled experiments, did not support it. For example, cutting tails of mice one generation after another did not yield tailless mice. This became known as the *central dogma* of genetics: the information flows from genotype to phenotype, not the other way around.[30]

There is no conceptual reason for heredity to rely only on the transmission of genes. The fact that this is how the DNA propagates results from the contingency of the development of life on Earth, and thus has little to do with universal biology. As Thompson notes, "[a]ny element of the developmental system that reliably recurs in each generation and that plays a role in constructing the evolved life cycle counts as something inherited."[31] Other means to pass down developmental resources have been discovered in the late twentieth century: for example, in the case of immunity developed to certain pathogens in mice, as well as in cases where viruses copy their own genetic information from host to host and can potentially carry organismic DNA with them.[32] The same applies to changes in the DNA as a result of exposure to ionizing radiation and certain chemicals.[33] Another way to pass on changes is through epigenetic inheritance.[34] Yet another option is hereditary symbiosis: organisms that live in another organism are passed to the offspring, e.g., as is the case with the microbial communities that inhabit termites' guts and make it possible for them to digest wood.[35]

It seems that the simpler the life form is, the more its heredity depends on the DNA, i.e., the central dogma-type inheritance and epigenetics. With symbiosis, for more complex organisms, we have an extra-genetic way of passing developmental resources down the line. Both of these are, however, contingent on the specifics of life on Earth. Yet if we are to consider increased sophistication that gives rise to organisms that can distinguish between the self and the world, perceive, and remember, we need to expect another level of inheritance, one that would rely on structures that are enabled by sensitivity and memory.

In more complex organisms and particularly in social animals, the transmission of developmental resources can be accomplished through social learning.[36] Here, behavioral patterns can be transferred to future generations by encouraging or discouraging certain behaviors, imitating observed examples, or in other similarly non-genetic ways. For example, if a certain canine had a troubling, fear-inducing, or painful experience with, say, cars, it might teach its offspring to run away when a car is encountered – for instance, by ushering them away at the sight of a car or even biting them if they do not run away quickly. If the repercussions impress them enough, chances are they will keep avoiding cars after they have a litter, and behave similarly toward their own pups, thus engendering a behavioral pattern transmitted through social learning from one generation to another.[37] On can also consider transferring certain skills this way, e.g., using tools for to catch insects in some groups of chimpanzees. In general, what we usually refer to as procedural knowledge, the *knowledge how*, cognitive contents that underly behavioral patterns yet do not require the ability to describe linguistically how they are carried out in order to demonstrate their mastery, can be transferred this way.[38]

The transferred behavioral pattern here is not encoded in genes, of course, but it is part of what one generation passes to another. As such, it also requires a physiological enabler: the ability of the recipient to acquire the behavioral pattern, e.g., by conditioning or imitation, and, of course, its ability to exercise it. Animals without a requisite nervous system that enables memory and basic learning would not be able to pass on any acquired behaviors.

In discursively intelligent humans another way of transferring acquired characteristics from one generation to another opens up: linguistic communication. Non-verbal animals are limited in their social learning by the inability to process declarative knowledge, sometimes referred to as the *knowledge what* – propositional knowledge, mental contents that state that something is the case and can be understood to state that something is the case by their carriers. To possess declarative knowledge that something is the case, the agent needs to address the relation between the content of its own mind, e.g., to assert such relation as true, as something that is the case, or as false, as something that is not the case. This requires the sort of metacognition that can address any mental content, including those that have no sensory component, a higher level of abstraction that cannot be supported by mere associative imagination – one needs such level to assert that something is true or false. This capacity is provided by language.[39] Shared language is also the only way to transmit such knowledge to another linguistic being, whether in the same generation or the next one, verbally or in writing. With the advent of the latter, the capacity for knowledge transmission becomes virtually unlimited in terms of the type of contents we might transfer – not merely behavioral sequences and skills but a wide range of meanings, from cultural traditions and aesthetical preferences to scientific knowledge and philosophy. It is also potentially unlimited in terms of the quantity of propositional statements to be transferred to the next generation. At this stage organisms can even think about editing the genetic code of other organisms and themselves, as we are witnessing nowadays.

Natural Selection

The third principle is Darwin's major contribution to biology: natural selection, also known as the survival of the fittest.[40] The principle was best formulated before Darwin's *Origin* by Patrick Matthew who did not consider it important enough to warrant a separate publication and tucked it in the appendix to his book *On Naval Timber*:

> [. . .] those individuals who possess not the requisite strength, swiftness, hardihood, or cunning, fall prematurely without reproducing – either a prey to their natural devourers, or sinking under disease, generally induced by want of nourishment, their place being occupied by the more perfect of their own kind, who are pressing on the means of subsistence.[41]

In other words, out of the interpersonal variety of all individuals of a generation only those would pass their genes to posterity who survive till the age they can reproduce and then actually reproduce. What happens to the organism after the reproductive age is not subject to natural selection in its classical, biological sense that considers its phenotype as a reflection of the genotype.[42]

Originally, the principle of natural selection was interpreted mostly as the survival of those who can win in the war of all against all: the strongest, the healthiest, the most cunning, the most aggressive. Reproductive capacities in plants and animals exceed the natural resources available for the offspring, so not everybody survives to pass on its genes. Therefore, individuals will have to compete with each other over resources – food, shelter, mates, etc., and the ones to survive will be those pines who are vigorous enough to grow taller and absorb the sunlight, those gazelles that are strong and swift enough to outrun most lions, those male lionesses that can catch enough gazelles to make it to the reproductive age, and those male lions that can fend off other male lions who are eyeing potential mates.

It is tempting to equate natural selection with the war of all against all given the vocabulary Matthew and Darwin used to describe it. It is equally hard to fail to notice other ways to survive, both possible to conceive of and those that are easily observable. Many plants and animals have little physical prowess yet still survive and multiply: camouflage that helps them hide from predators, having great many seeds spread around, being parasites on other plants and animals are but few examples of adaptations that have little to do with aggressive performance. Perhaps, the social and geopolitical conditions during Darwin's times impacted his understanding of natural selection: the British Empire was on the rise, conquering new lands and frequently rubbing shoulders with other colonial powers in a way that was far from friendly; the colonial project was in vogue and thus provided a thinking paradigm for comprehending other areas of knowledge, including biology.[43]

Several less militant survival strategies have been suggested. Of a particular interest is Kropotkin's proposal that mutual aid between animals, at least those of the same species, can be as important factor of evolution as the struggle between them.[44] He arrived at this conclusion after observing animals living in the harsh conditions of Eastern Siberia and Northern Manchuria. There, for elk and rodents and many birds, sticking together and sharing resources like food or shelter might be a better strategy for survival than fighting each other. In fact, we can see similar phenomena around the world: in elephant herds, wolf packs, mole rat colonies, etc. In addition, few animals who care for their offspring ever compete with it: in this case, competition would preclude survival.

One can think of other strategies as well. For example, much has been said about the peacock's magnificent tail. It puzzled Darwin quite a bit: after all, what can be the evolutionary benefit of such a wasteful and apparently useless trait that also makes the peacock more conspicuous and the chances that it will be eaten by a tiger – much higher in comparison to a humbler colored peahen. Yet this dangerous conspicuousness might be pivotal to its contribution to the survival of the peacock species: the role of the peacock may well be to mate and then to attract predators to his overly decorated self, thus increasing peahen's and chicks' chances of survival. Little can be expected from a peacock in terms of parenting and protection of its offspring, given its humble mental abilities and little physical prowess. The number of peacocks required for sustaining the species is smaller than the number of peahens. Given these, for a peacock being a bait to detract tiger's attention from the peahen and the chicks can be a decent evolutionary strategy.[45]

To summarize, as Dobzhansky noted, "natural selection is neither egoistic nor altruistic; it is opportunistic."[46] Due to natural selection the genotypes that produce phenotypes fit for their environment persist; and there can be more than one fit genotype. Yet in the long run cooperation seems a more important trait, as it produces more stability, reduces confrontation between individuals within a species and thus the chances they will not make it to the age of procreation. As a support for this claim, Dobzhansky notes that for a parasite to become somehow useful for its host means that the host will at least stop fighting it.[47] Yet when we consider cooperation between the members of the same species, we would see that it can lead to better defenses, to increased efficiency in getting food, and to other similar benefits that all favor survival. Such cooperation, in turn, will become a factor in natural selection: those individuals who are good at cooperating will be favored in the survival game, and with them – characteristics that make them better cooperators. For example, ability to communicate with other species members is of a limited use for solitary animals: its only purpose would be advertising themselves for potential mates. For animals who live in groups, on the other hand, ability to produce signals and respond to signals would be very beneficial, as the ability to create and learn to register new signals. Physiological structures, including those of the nervous system, that enable communication will be thus favored for cooperating animals by natural selection.

One can think of natural selection as a set of external pressures that the organism confronts in its environment. However, organisms are embedded in their environments, and their actions matter for the environmental capacity to support them and to the kind of pressures they experience. Even in the simplest organisms, the way they consume nutrients and the way their waste can be absorbed impact the environment and the support it can provide for them. The more sophisticated the organism is, the more its impact on the environment is important for its own evolution. Motile animals can move to reach food and even migrate; social animals' environment includes other animals of the same species; discursively intelligent humans' impact on the environment has no principal limit. This way, "organism and environment construct each other in development and evolution,"[48] and we can talk about co-determination of organism and its environment. This would be particularly important when we consider individual organism's self having its wants pursued in its world.[49]

Most importantly for our project, the capacity for choice is also subject to the pressures of natural selection, as well as its enablers: sensations, central control over the body, representation, and memory. This is a direct consequence of organism's embeddedness on the world, where the activity of the organism, behaviors originating in the living entity, impact its interaction with its environment and this way – its survival. I will argue later that this capacity, once it appears, also changes the rules at play in natural selection, and when we consider human freedom, changes its character completely.[50]

Increase in Complexity

Lamarck is mostly known for what came to be called the *Lamarckian inheritance*, his suggestion that characteristics acquired by animals prior to giving birth can pass to their offspring, as it was noted earlier. However, his main argument was that life inevitably evolves toward greater complexity. If all living bodies are products of nature, then, argues Lamarck, it would make sense to believe that they were not produced all at once. And if they evolve, then it makes perfect sense to start with the simplest. If not for the increase in complexity, taking place through reproduction that preserves acquired modifications in mortal living beings, the nature would never be able to produce so many species. To support his argument further, Lamarck notes that the highest faculties could not be created at once, as "they are found only in conjunction with highly complex systems of organs" which are a result of many eons of evolutionary development.[51]

Lamarck's argument hinges on the assumption that nature always starts with the simplest form of life; for him that was the constantly happening spontaneous generation. The hypothesis of the simplest organic form born out of the chance combination of chemicals that are present in inanimate nature has the support of both the consideration of probability and the absence of other sources barring divine intervention.[52] Yet the necessity of the increase in complexity and the eventual appearance of discursive mind has been justifiably doubted. We cannot suspect natural selection of having any *telos*, and it advances through the myriad of changing environmental factors that test organisms' fitness for the particular environmental configuration at the particular point in time. Therefore, one is justified in arguing that every organism and every feature, including *homo sapiens* with its discursive mind that greatly appreciates itself, are a result of a string of contingencies, and the evolution of life could have proceeded differently. What remains, then, is to explain the development that has actually occurred, i.e., the evolution of a person who is contemplating evolution.[53]

Explaining the developments that have already occurred can lead to some fruitful insights. For example, one can note that the "very fact that every new species variant builds upon a pre-existing life form suggests that evolution has a cumulative development that necessarily advances to more and more complex organisms."[54] It is hard to imagine multi-cellular organisms arising without their unicellular evolutionary ancestors, the emergence of animal motility without plant nutrition, and memory without sensitivity. These considerations provide a philosophical ground for understanding the evolution that has occurred. They do not, however, provide a necessity of evolutionary advancement. What they show is that it is necessary for the less perfect[55] life forms to exist for the more perfect life forms to evolve, yet they do not demonstrate the necessity of this evolutionary direction: "[t]o the extent that the blind contingencies of natural selection determine which new species emerge, whether increasing complexity results is never secure."[56]

There is, however, another way to consider evolution in the context of complexity. For the sake of this discussion, I will treat complexity of a living entity as having more parts and more relations between them, as well as more functions it can perform, including psychological functions.[57] With that, there are three possible ways to cope with the challenges of survival: evolve to be simpler, evolve to be more complex, and stay the same in terms of complexity while developing new features instead of the old ones.[58] There is no principal limit on or a reason to prefer one of these three evolutionary strategies. Under these conditions, there always will be organisms that will evolve toward more complexity, due to sheer chance provided by the mechanisms of variation. And, more importantly, the survival chances of the more complex organisms frequently will be better. Moreover, the improvement in their chances will grow exponentially and reach its so-far-known peak with discursively intelligent humans. This, because the survival pressures at the top of the available complexity scale will be lower.

The movement of the gas molecules in any container is random, yet, given a level of pressure that is higher inside the container in comparison to the level of pressure outside of it, these molecules will start escaping through any opening – as opposed to the situation of equal pressure, where the movement of gas from the container out will be quantitatively the same over time as the movement of the gas outside or the container into it. In the first case, the randomness of the individual movement of the molecules causes specific directionality of their overall movement: it creates the level of pressure that dictates the direction. With evolution, we can see a similar effect. If organisms evolve to possess less complexity or keep it at the same level, they will keep competing with other organisms occupying their ecological niche. Even if the new features enable them to conquer their corner of the Earth at the expense of other species, when they multiply, the competition will be back to the previously experienced levels. By exploring new niches, e.g., mammals getting back to the sea, this effect only gets delayed in time. On the other hand, with the increase in complexity animals get to a new level, they now face less competition even if they keep living in the same environment. The clearest example would be the transition from the plant to the animal life form. By acquiring motility, living beings can now move around and get access to new food sources rather than waiting for them to flow toward their roots. This niche is still unoccupied, and they can thrive. Once there are more animals, acquiring keener perception – an increase in complexity – is a winning evolutionary move.

At a certain point, improvements in organism's ability to control its self lead to the appearance of memory, and this is a leap in the ability to survive: remembering past experiences and associating them would greatly increase the ability to extract food from the environment and escape danger. Less sophisticated animals, on the other hand, would fall prey to the same competitors whom they barely escaped last time if they have no recollection of the previous encounter at the sight of danger, would not be able to repeat the previous action that, attempted without much planning, led to extracting food – they simply would not remember. Yet the crucial ability that come with mind is the capacity to cope with change: try something, remember it, and then try again.

The very same pressure that we witnessed at earlier stages, however, will soon build up, with the multiplication of competitors that possess memory, the ability to associate images and generalize over them, and more. Then, mechanisms of passing acquired information to others and to posterity will become important, as they provide a clear survival advantage: those who are quickly informed about the approach of predators and about new sources of food have a better chance at survive; the ability to train the offspring in hunting or escaping danger avoids the necessity to wait for the appropriate change in the genotype that could take many generations. This, in turn, changes the structure of the evolutionary pressures and favors more sophisticated cognitive endowment.[59]

With a capable nervous system that can remember, associate memories and sense data, interpret signals from other animals by associating them with generalized remembered images, the little step can be taken toward a mind that can relate to the relations between its own contents, go meta – a discursive mind that can give rise to language. This mind is disproportionally more adaptive than the non-verbal animal's mind: it can envision radically new solutions to challenges, not only produce images based on the associations already made between the encountered ones. Language is crucial for this, as it enables assigning words to any kind of mental content and manipulating such contents through these words. Without language, an animal capable of retaining sensory images and generalizing over them can envision itself using a prolonged object reaching out for food: it has experience with objects of various lengths, with food which is placed high up, with its own movements, etc., and can generalize over these images. Yet to envision cultivating bananas one needs to think of the process of plant growth and a way to relate oneself to it. This is no mere generalization of images – plant growth is an abstract concept, as well as one's relation to growing plants. As the previous sentence demonstrates, we need names to address it – and it is language that supplies them.[60] The move toward a mind capable of discursive metacognition might be accidental, yet its survival is not – just like with all the previous cases.[61]

This way, randomly acquired improvements have a directionality toward greater complexity, directed by the pressures from evolutionary peers. The next section will explore it in the context of Hegel's thought, where each new evolutionary stage is the truth of the previous one. Later, the evolutionary direction toward complexity will be discussed as leading to the increasing levels of freedom in Chapter 5.

HEGELIAN EVOLUTION

Hegel opposed the type of theories of evolution that have been known at his time, namely the Lamarckian evolution. In the frequently cited passage from the *Philosophy of Nature*, he addresses theories that suggest the emergence of "plants, polyps, molluscs, and finally fishes" from the water, land animals – from fishes, and finally humans from land animals, and suggests that the "land animal did not develop *naturally* out of the aquatic animal, nor did it fly into the air on leaving the water, nor did perhaps the bird again fall back to earth."[62] Hegel contraposes this type of theories with the emanation theories that explain the diversity of life on Earth by the degrees of degradation of God's image – and finds fault with both.[63] One might find this odd. Hegel's system is dynamic in all its aspects and suggests the development of Spirit – so why not evolution?

It seems that the reason of Hegel's dissatisfaction with his contemporary evolutionary theories is the same for which he rejects the theories of emanation: they do not provide a conceptual, theoretical explanation for the diversity of life on Earth, they merely try to develop a theory that explains data. While this might be acceptable for empirical sciences, it falls short of what reason would require.

However, one can think of a Hegelian evolution – a conceptual framework that would explain why life is bound to evolve, as Jonas starts doing in his *Philosophical Aspects of Darwinism*[64] and Winfield develops further in *Universal Biology*.[65] Such philosophical, universal theory of evolution would explain not only why life necessarily evolves but will also make an attempt at charting the course of development that might be suggested by the very nature of life. In other words, this kind of theory will need to explain *why* it is the case that "[a]nimal nature is the truth of vegetable nature"[66] rather than otherwise, why would the development of the organisms possessing a certain nature give rise to organisms of a different nature, how the processes at work in any possible environment would be conducive to this course of phylogenetic development – or, failing to do so, admit defeat. Such monumental undertaking exceeds the limits of the current chapter. Therefore, it will be pursued here only to the extent that is necessary for clarifying how freedom evolves.

What is the driving force behind evolutionary development? Darwin, being a naturalist, suggested that natural selection drives evolution, as it has been described earlier. Those of his contemporaries who were not frightened at the prospect of having common ancestry with apes were struck with the obviousness of this suggestion. This obviousness makes it a great candidate for philosophy, and specifically for thinking about a philosophical category that would be inclusive of it. Such a category, thought about in Hegelian terms, would address a universal force that is able to explain why the truth of one life form is another life form, and not just why the fittest survive.

I suggest that this category is lack, a sub-species of contradiction. Lack breaks the equilibrium, and thus provides an impetus for change, for becoming. It is this "unrest of something in its limit in which it is immanent, an unrest which is the *contradiction* which impels the something out beyond itself"[67] that drives evolutionary development. The contradiction between pure being and pure nothingness that makes one to realize their essential sameness, and thus proceed to becoming. Similarly, the contradiction between the need and the current state of affairs pushes the subject to make a move, e.g., a hungry goat to search for food. On the evolutionary plane, it would be the contradiction between the abilities of the organisms carrying their respective genetic endowments, similar in most respects yet different in some, to function in a given environment and their needs as living beings that would cause certain organisms to survive and pass their genetic code further, and others to perish. The lack of adjustment is the category under which natural selection labors, where the meaning of adjustment changes from one entity to another and from one level of capacity to the next: surviving when the water dries up for three months, being able to stay warm when the weather changes, fulfilling the emotional need for love, or satisfying the culturally engendered need for personal freedom are all examples of overcoming a lack.

At the level of the specific organism lack can be seen most clearly. I suggest that such lack exists in all organisms: any environment, be it physical, social, or cultural, poses challenges that the organism needs to cope with. Some such challenges come from the availability of nutrients, some – from other organisms. These factors work together as well: plenty of nutrition causes organisms to multiply and consume much of it, and thus creates scarcity of food; this works for all life forms that cannot consciously create new sources of food, plants and animals alike. Lack is a constant companion of life from its very inception, and hence natural selection.

There are two ways to cope with lack: to change or to die. Furthermore, there are two principal ways of philosophically interesting change: becoming simpler or becoming more complex.[68] This way, a species can develop the ability to digest more types of food in response to the lack of the one it had been used to before it became scarce due to climate change, population surge, or appearance of competing organisms. It can be achieved, for example, by organism's digestive system becoming more sophisticated, e.g., by secreting digestive juices. This, by having the organisms that have the germ of this ability survive, and those that do not – die out. Alternatively, given favorable environmental circumstances, the species can lose the complexity of its digestive system: it requires energy, and if the individuals with a genetic makeup that gives rise to a more primitive digestive system can survive, they would have a reproductive advantage over those who need more food to maintain their useless complexity. Anteaters are a good example: their digestive system is rather simple, as the ants they consume provide enough acids to break down their own bodies. The third options would be developing abilities that do not constitute an increase or a decrease in complexity: the foray of mammals into the ocean comes to mind. This last option, however, does not seem to be interesting philosophically.

This way, the truth of an organism that is lacking *qua* organism trying to cope with its environment, the resolution of its conflict would be another organism. It can be simpler, more complex, or similar in terms of complexity. Yet when we talk about life forms, categories of the type of life, the truth would always be a more complex life form.[69] This, because the contradiction at the level of a form of life cannot be resolved by simple adaptations. The lack of plants, for example, is a contradiction that is not just incidental, contingent on physical conditions that might change, but inherent in their immobile, passive life form that requires constant flow of nutrients from their immediate environment. Every time the immediate environment of a plant changes, this lack shows itself, as it is rooted in its life form's inability to reach out to nutrients that are farther away. This is a major evolutionary disadvantage that is detrimental to survival. The overcoming of this disadvantage requires acquiring an ability to move toward the sources of nutrition, thus giving rise to a new type of metabolism. Yet this new function of motility which is necessarily bound with more advanced abilities, namely sensitivity and central control over the organism, requires much greater sophistication. Similarly, to overcome the inherent deficiency of the animal life form, its dependence on the physical environment as it exists within the animal's reach, what is required is the ability to construct a new environment, not merely to change this or that aspect of it or the ability to travel a bit farther. This demands a much greater complexity in the very mechanism that was born with animal life – the mind, an outgrowth of organism's central control.

The overcoming of lack inherent in a life form type by the way of evolving into a more complex life form is bound with the higher degree of freedom, self-determined choice. Chapter 5 outlines the course of this development, yet first to address briefly the question of the additive vs. transformative nature of biological evolution.

BIOLOGICAL EVOLUTION: ADDITIVE OR TRANSFORMATIVE?

Aristotle viewed the hierarchy of life forms as additive: animals add perception, desire, and motility to the nutritive life of plants, and rational animals inherit all these characteristics, adding reason on top of them. It is tempting to conceive of biological evolution on similar terms, where "each mutation adds to the cumulative endowment of prior evolutionary developments."[70] This seems to be the case at two levels: the biological micro level and the conceptual macro level; however, at the level of behavior and the functioning of physiological structures, the level most relevant to choice and freedom, the conception of evolution as additive is in need of further refinement.

At the micro level, one can point out to the great commonalities of human, non-verbal animals', and plant DNA, where each higher life form seems to add new genetic sequences while preserving most of the old ones.[71] At the macro level, we can point to the classes of physiological functions and behaviors that Aristotle noted: all living things consume nutrients; animals consume nutrients but also perceive, have urges, and move; humans consume nutrients, perceive, have urges, move, and think about evolution. And yet the actual ways of functioning, for example, the way oaks and dogs and financial advisors consume nutrients, differ greatly between different life forms. Oaks absorb whatever the contiguous environment offers, given that their fixed physiological structures are receptive to it. Dogs not only move to get to their food but also can refuse it if they associate it with danger or dissatisfaction of their significant humans – at least well-behaved dogs, that is. Financial advisors might refrain from consuming food not merely on emotional but also on ethical grounds, for example, if they think that it is immoral to kill animals in order to eat them, if they consider the food they have access to ritually impure, or for many other reasons that have nothing to do with accessibility and sensation. Yet the difference goes deeper: all elements of feeding behavior are pervaded by the character of the life form. Emotions arise following ethical positions, physiological reactions like and increase in the heartbeat frequently follow, and humans can even intentionally end their lives for causes that have to do with what they think to be the case. The character of a life form is not a result of merely adding another layer to the ones characteristic of the life forms that precede it in the evolutionary development – it results from transforming the existing elements by the new addition. Similarly to Agamben's notion of the living self using itself,[72] each life form utilizes the physiological resources it inherits from its evolutionary predecessors and makes a new use of them, thus creating a set of new behaviors. Without this understanding it would be hard to explain how relatively small genetic changes give rise to organisms with very different capabilities: these additions do not create radically new organs or organismic systems but alter the use of the existing ones. These slight genetic shifts, if proven useful, make the organism subject to a new set of environmental pressures which, in turn, constitute an environment that gives advantage to certain mutations, and those are the ones that get passed to posterity.

For example, *trichoplax adhaerens* resembles simpler plants in pretty much everything. It absorbs nutrients that come into direct contact with the underside of its one millimeter of a body, does not have differentiated organs, and reproduces asexually, by budding. However, it is not rooted to one place and possesses a rudimentary capacity for movement using its cilia, organelles found in eukaryotic cells that *trichoplax* happens to have on the exterior of its body. This minimal difference between this proto-animal and its immobile kin opens before it a whole new way of life: now it acquires food by moving around, thus changing the nature of its metabolism – it is now mediated, even though in the most primitive form. This new way of life exposes it to the new set of evolutionary pressures: now those mutations of *trichoplax* that contribute, even indirectly, to its ability to move faster and farther would have an evolutionary advantage, and thus have a chance of being preserved phylogenetically; the same mutations would be completely useless to immobile plants. All its other capabilities will be now subjected to the pressures of natural selection in light of its motility. For a plant, the ability to register suitable nutrient at a distance is useless: it does not move. Yet for the evolutionary posterity of *trichoplax* it would be extremely useful, thus creating a pressure toward the development of perception.

Similarly, apes have pretty much all what is needed for human functioning: well-developed and well-coordinated limbs, keen vision, capacity for social relations, and more. What they lack genetically is a neurophysiological basis slightly more developed than their own, a bit more of the brain processing capacity that would enable them to go beyond generalized associations. In fact, it well might be that some individual chimps get very close to the humanlike capacity here. With that slight change, though, acquired by human ancestors, the structure of evolutionary pressures changes dramatically: the new capacity makes a different use of pretty much every organ of the body. Brains now can be used for planning for the future and thinking up romantic stories, hands for making sophisticated tools and decorating pots and writing, legs for kicking balls, and more. These cannot be achieved through general representation. Romance is not about the feeling of lust, it is defined by a conception of attraction that is different from a physiological one; as such, it is impossible to arrive at by generalizing over the sensory imagery. Similarly, inventing a bow is beyond generalizing over the images of sticks and strings – it requires judging different ways of putting together materials of varying degrees of flexibility against the criterion of distance and force of impact. Finally, the game of soccer is an abode or rules that have normative character – something that is unreachable for a mind that does not get beyond generalizing sensation. The conceptions of themselves that humans develop have a great impact on the way the body is treated: now it is covered in cloth not only to protect it from the elements but out of modesty; the conception of the behaviors and bodily shapes appropriate for different sexes becomes a factor in choosing mates; and more.

Those slight changes that push organisms to a new evolutionary plateau with its own set of factors that impact survival can be seen as underlying the main philosophically interesting evolutionary transitions: from unicellular to multicellular life, the emergence of plants, the emergence of animals, and the appearance of discursively intelligent agents, in our world – humans. Each of these changes radically alters the type of the life form: the old capacities acquire new meanings for the survival of the organism and frequently take on new roles; all this due to the changes in organism's mode of existing.

Analyzing the uniqueness of humans, Heidegger in his *Letter on Humanism* argues that our existence is so radically different from the existence of other living entities that it deserves a different name: ek-sistence (*Ek-sistenz*).[73] This is because humans engage with truth while non-verbal animals are oblivious to it – humans possess the capacity to represent things to themselves, for example, represent beings as beings,[74] and contemplate the representation; evaluate truth and falsity and discuss the meaning of truth and falsity. All this brings about a radical break from the animal life form.

Heidegger thought that this radical difference is unique to humans. However, when we consider the transition from plants to animals, we see a similarly deep divide. The existence of the animal life form is at least as radically different from that of plants as the human one is from that of non-verbal animals. Given the variety of animal life it is hard to see, yet when we consider, say, hyenas and pine trees, both fairly typical examples of their respective biological kingdoms, we would find that their way of life and evolutionary pressures are radically different, while the differences between pines and moss are not so striking, as well as the differences between hyenas and octopuses.

Heidegger sees the uniqueness of the discursively intelligent life form as rooted in its ability to think, which is rooted in language that, he argued, "houses ek-sistence."[75].However, if we are to account for the radical differences between all life forms, we might turn to the question of freedom.

It is the relation to the potential for freedom, to the ability to evolve self-determined choice, that makes for the abyss between plants and animals. The plant life form cannot develop freedom: its self, immobile, non-sentient, lacking centralized control over itself, cannot choose between alternatives: it can neither register them nor move itself to act to choose one. Plants do not act, they do not behave; however, animals do. With the advent of the animal life form, the stride toward freedom begins. Even simpler animals have the physiological structures that can be used to choose. With these structures, specifically those related to movement and sensitivity, natural selection would favor those who can exercise them in a coordinated fashion: this is the way animals acquire nutrition and procreate. Thus, structures that control the moving and sensing organism develop further, acquiring, as it has been described in Chapter 2, progressively more sophisticated associative memory and finally – the ability to choose. Not all animals are capable of choice,[76] yet it is the animal life form that can evolve toward it. Self-determined choice, in turn, changes the landscape of the evolutionary pressures – the same landscape that will be changed again with the development of language; as I will argue later, this is a development to which the ability to choose leads.

Chapter 5

Evolution as the Unfolding of Freedom

STARTING DOWN THE EVOLUTIONARY ROAD

Conditions for the Emergence of Life

The question of evolution is independent of the question of the origin of life. Living entities evolve, no matter how they or their phylogenetic ancestors originated. In principle, it is possible even to simulate the conditions in which life can appear in a lab; serious steps toward this have been undertaken since the famous Miller–Urey experiment in 1951, where organic compounds necessary for the life of Earth as we know it, and specifically amino acids, were generated in a relatively simple setup including evaporating water, methane, ammonia, hydrogen, and electrical sparks simulating lightning.[1] Similarly, there is no necessity that life emerges at all: inorganic matter would give rise to living entities only through a serendipitous combination of conditions that happens to take place. The specific circumstances of the planet Earth led to the appearance of DNA-based cellular life that relies on specific chemical compounds, yet it is possible that in another place and in another time life could have been based on a different set of chemical elements. Similarly, nothing prevents future emergence of a form of life that is differently chemically instantiated.

With that, the specific conditions under which life emerges play a crucially important role in evolution: they limit the biochemical composition of the forms of life that can evolve by providing the biological starting point and the initial set of conditions which impact the survival of the organisms. This way they shape the evolving life forms. The conditions on the 2.3 billion years old Earth led to the abundance of ancient cyanobacteria, the ancestors of today's blue-green algae. The cyanobacteria, in turn, most likely led to the abundance of oxygen in the 2.3 billion years old Earth's atmosphere.[2] The availability of oxygen on Earth and the rarity of other compounds provided a ready path for the development of the aerobic life forms and thus impacted the course of evolution – which, in the case of, say, scarcity of oxygen and abundance of chemically inert argon would have proceeded differently. Thus, we have to consider the interplay between the environment and the specifications of life as a framework that impacts the initial evolution and later – the alternatives for choice free living entities can access, as well as those they can conceive of and develop.

Therefore, we have to consider the "minimal specification of the processes fundamental to life"[3] in the context of the ancient biosphere, the inorganic environment that has the potential to give rise to and then host living entities.[4] As it was established earlier,[5] at a minimum living entities are characterized by metabolism and organic unity; a form of life that is to have a chance of continued existence is also characterized by reproduction. Even the simplest organisms absorb materials from their environment – in order to do work, i.e., to maintain any life process, they have to expend energy, and the first law of thermodynamics, that of the preservation of energy, applies to them as to all other physical entities.[6] Living entities ought to maintain organic unity, where all parts of the organism, however simple, work in unison. Finally, living entities reproduce, giving rise to new living beings of the same kind.

Life has to emerge from the inorganic nature, simply because this would be the only type of matter available for it to emerge from; this has been recognized by thinkers as diverse in terms of their focus as Hegel and Lamarck.[7] Today other theories are being argued for and tested, yet the initial emergence of live from non-life is hard to doubt. What is interesting, though, in the context of evolution is the character of the processes that give rise to life, as they play an important role in defining the parameters of its development. Such processes cannot be a mere rearrangement of parts of matter in space: metabolism requires changes in matter, not just in its arrangement, as through metabolism the organism takes matter from the outside and turns it into its own components. Therefore, life has to be based on chemical processes, processes that alter the character of the materials they start with and produce new ones.[8]

Life's Chemism and the Cell

Hegel argues that life is a combination of chemical processes that renews itself[9] – what constitutes life is a bunch of chemical processes that spontaneously renew their activity, perpetuate themselves. This self-renewing, self-sustaining nature distinguishes between the non-organic and the organic.

Non-organic chemical processes are caused by external conditions, i.e., an arrangement of chemical elements in a close enough proximity and under the conditions of proper temperature, atmospheric pressure, etc., that are necessary for the chemical reaction to occur. Once the supply of the reactants is exhausted, the process stops transforming them into the products of the chemical reaction. Yet once several chemical processes are aligned in a way that they renew their constituents, e.g., by continuing to absorb them from the outside, and themselves, they become organic and give rise to life:[10] a metabolizing entity that absorbs the ingredients needed for its chemical processes, where parts of this entity, i.e., whatever maintains the processes in their reciprocity, work together, and the entity is capable of rebuilding its own structure that will maintain similar chemical processes. This self-renewal is usually referred to as *autopoiesis*.[11]

How can this fortuitous co-occurrence of chemical processes take place? Firstly, the requisite chemical components that are subject to chemical reactions capable of sustaining each other need to be present. This can be achieved through the processes simulated in the Miller–Urey experiment mentioned earlier.[12] Secondly, these components need to elicit the requisite reactions that will result in the structures necessary for the sustaining of the very reactions that gave rise to them. These structures might develop by piggybacking on inorganic materials that tend to replicate under certain conditions in specific patters. Cairns-Smith, for example, suggested that clays can be such materials: they form crystals the patterns of which repeat when exposed to certain environmental conditions, and their surface tends to attract other chemical components.[13] If such components are right for sustaining self-renewing chemical reactions, all what remains is a chance for them to get detached from their silicate base and get enclosed into "boxes" formed by other molecules that, due to their physico-chemical properties, tend to form enclosures – and we get a proto-cell with a molecule inside of it that can be, for example, a relatively short RNA that is "known to do pretty much everything life does – reproduce, mutate, adapt, use energy, catalyze reactions."[14] This or that way, it is possible to conceive of the formation of structures that are basic to life.

Life might arise by chance. Yet once it emerges, "it becomes unconditioned or self-conditioned"[15] – it depends on its environment to have the elements that the chemical processes biologically constituent of life need to maintain themselves. The emergent self, the bundle of structures that support such processes and are constantly re-created by them, is self-active. It creates its own conditions and interacts with its environment on the terms encoded in its very structure, maintains its own identity "in and through its own life process."[16] As such, it needs to possess several characteristics in order to be capable of evolving.

Evolving Biological Entities

In order to evolve, an entity needs several crucial features. Living beings should have their own individual selves, be biologically separate from other living individuals, so each can be a subject to the pressures of natural selection. They have to be in need of various elements of their environment – otherwise organisms will not experience any pressures to change: having no needs at all would mean experiencing no lack, being subject to no constraints on survival, a biological nirvana that seems impossible for living beings and certainly not compatible with evolution. They should have a world, be responsive to at least some of the features of their environment; this, in order to metabolize and to feel the pressures of natural selection. They also should be connected to the world, be alive in the world in a way that is conducive to satisfying their needs – this connection should link the self, the world, and need in a way that allows for natural selection to work and for at least some organisms to survive and give rise to their offspring. For organisms that actively seek satisfaction for their needs, such a link should be an active position toward the world, care for their position in the world. Finally, living beings have to be plastic, to be capable of change, at least across generations – without that evolution is impossible. These aspects will be briefly analyzed below as individuation, need, world, care, and plasticity. Then, I will describe the main evolutionary developments – the appearance of multicellular, plant, animal, and discursive life forms – in terms of the changes in these aspects, while arguing that these developments enable the emergence of freedom and, once it emerges, are impacted by its unfolding.

Individuation

Life is an autopoietic entity, self-making being with its own self-centered teleology, its for-its-own-sake intrinsic purposiveness.[17] The aspect of autopoiesis that is particularly important to the evolution of freedom is the biological separation of living organisms from their surroundings. While living organisms should be open to their environment to obtain nutrients, having their own inherent purposiveness requires organic structures that would support the organismic self. Such structures, as defining the biological infrastructure of organism's self, would necessarily be separate from the environment, not contiguous with it. In each living organism, we would see some sort of instantiation of these structures. For example, in the smallest form of life on Earth, unicellular organisms, we would have DNA, whether floating around the interior of the cell in prokaryotes or enclosed in a nucleus in eukaryotes, encoding the main parameters of the processes that are needed to sustain the physical existence of the organism. Together with an adequate environment that provides the thermal and other conditions and the chemical elements necessary for the life-sustaining processes to take place, they define the character of the unicellular organism, as the genes belong to the cell and are located in the cell. The cell membrane ensures the separation between the organism and the environment in such a way that the organism itself regulates the intake or nutrients and the expelling of waste – the commerce with the environment is done on organism's own terms. For other life forms, other ways of preserving uniqueness and separation would be in place, including psychological, social, and cultural. For example, remembered images are an important source of individuation for organisms that possess memory: these are the images that impact much of organism's behavior when its sensations are compared to them. Discursively intelligent humans hold conceptions of themselves which include not only memories but also beliefs about who they are, much of which are culturally induced and socially impacted. These and other sources of individuation will be discussed later for each life form type.

Individuation is crucial for evolution in general and for the evolution of freedom in particular. If not for the individuation, it would be impossible for organisms to be unique. If organisms are all uniform, natural selection has nothing to choose from, and the very moment the conditions are not right for one of them to exist, all will perish. As it was noted earlier,[18] each organism constitutes a hypothesis to be tested for making it to the reproductive age and then multiplying – and for that it should be unique at least in some respect that is relevant for survival.

As for the evolution of freedom, organism's individuation is a necessary condition for self-determined choice: without an individual self, there is no choice carried out by the self – these are the aspects of self-purposiveness and organism's own terms that constitute the basis on which any notion of choice can ever develop. Specific biological mechanisms that are involved in individuation will also be involved in choice in several ways. The development of biological and psychological features relevant to choosing is rooted in the options each stage of evolutionary development makes available to the next stage. For example, individuation by a semi-permeable membrane that encloses cellular material, allows expelling some molecules, and responds differently to various chemicals, provides a structure that can later be used for communication, as it is the case in slime mold.[19] The biochemical capacity underpinning cellular communication might provide means for animal signaling, e.g., chemical communication in ants. In more complex animals, the structures supporting communication create neural pathways that later can be utilized to support other types of signaling, and then – the physiological basis for the distinction between subject and object and language-based communications. These structures also evolve into ways by which alternatives of choice are made accessible to choosing individuals: after all, for a ram to find a ewe ripe for mating, it needs to see and smell it.

Need

Life entails constant expenditure of energy needed for its biochemical reactions. Such energy is acquired through the metabolic process whereby organisms take in nutrients. This means that life is ever needful: having needs, at the very basic level, the need for nutrients, is inherent to life. This feature of life is rarely addressed explicitly, and if it is, it is handled as part of the of the discussion of metabolism[20] or in the context of freedom – as a limiting influence of life's physique on choice.[21] Yet the fact that need is a feature of life, any life, has serious significance for the evolutionary process in general and for the evolution of freedom specifically.

A need can be satisfied or nor satisfied. In the case of living entities, it is highly likely that at various stages of its life cycle an organism will have some of its needs not fully satisfied; unless the satisfaction of its needs is provided for by an external agent who understands these needs perfectly and is willing and capable of catering to them – a condition that can be created, perhaps, in a lab in the case of a very simple life form, yet would seldom be present elsewhere. This is because living entities are individuals separate from the environment in a way where their metabolic and other processes are not coordinated with those of their surroundings. The simplest unicellular life form, if presented with the conditions where nutrients are abound, will multiply to the extent where the nutrients would not be sufficient to sustain the whole population – and then experience lack. With more complex life forms, e.g., multicellular plants that have multiple parts, it is hard to imagine that water, nutrients present in the soil, and sunshine will always be present in sufficient quantities, such that no competition from other members of the vegetable kingdom would ever strain them and thus cause the lack of resources, at least temporary. With animals whose metabolism is mediated, lack is experienced most frequently, as hunger; yet we can also think of the lack of satisfaction underlying lust, or lack of satisfaction of emotional needs, e.g., love and affection. In humans, the domains where lack can be experienced stretch as far as the outer reaches of the scope of human needs, including intellectual needs. All these will be addressed when different life forms will be discussed, since lack, the dissatisfaction of needs, is the driving force of evolution, as I argued earlier.[22]

There are different ways to alleviate lack. One, exemplified by the plant life form, is to evolve to minimize energy expenditure while absorbing nutrients constantly. Another, represented by the animals, is to develop psychophysiological structures that push the organism to see the satisfaction of its needs – desires. Desires are necessary to bridge the "recurring gap between unfulfilled need and its satisfaction"[23] – they are not merely allowed by such a gap, they are necessary for the survival of a life form that is characterized by this gap. These are the desires that push the animal to act and sustain its action over the time required to get food, escape danger, or consummate mating – to bridge the said gap. It is through desires that lack becomes the most effective driving force of evolution. The neural equipment that provides the physiological infrastructure for maintaining and acting upon desires becomes an evolutionary advantage: animals that possess it are more capable of bridging the gap between the needs they have and their satisfaction. Emotions develop: these are *practical* feelings, ones that push the animal toward an end.[24] The same neural structures work together and thus give evolutionary advantage to perception and memory that enable carrying out desires, and thus lead us toward the possibility of choice. This phylogenetic development, of course, relies on contingent factors, yet its vector seems clear: once we have a need, and variation produces simpler and more complex individual organisms alike, the presence of need that goes hand in hand with lack drives evolution toward animal choice and eventually human freedom.[25]

World

For every physical entity one can define a surrounding – the set of elements that can impact the said entity and the set of elements that might be impacted by it. This definition is spatio-temporal: as laws of physics apply to all material entities, we can talk about one physical world that includes all physical objects.[26] Living beings, on the other hand, constitute selves that are meaningfully separate from their physical surroundings, as they carry their own developmental and behavioral agenda within their bodily boundaries and interact with what is around them on their own terms. Therefore, the surrounding of the living beings is *their* environment comprised of the elements that carry significance for them, as opposed to other elements to which they are oblivious: the latter are not allowed in by the amoeba's semi-permeable membrane, are not absorbed by the roots of the tree as nutrients, are not registered by the dog's sense of smell or vision, are not part of the human knowledge if humans do not have the means to know it or have not applied adequate means yet to this end. What aspects of the surrounding are meaningful for the living entity is defined by its way of life. This way, as von Uexküll notes, using the term *Umwelt* to designate the living being's environment, for the biologist there are as many Umwelts as there are living beings, and these worlds are intelligible only in connection with these types of organisms.[27]

The Umwelt of an organism consists of the sum total of those aspects of its surrounding with which the said organism can interact. The stimuli that are registered by the organism and elicit its response, or indicators, are its environment-as-registered, or its *perception world* (*Merkwelt*); the activity of the organism in those aspects of the world that it can impact – its world of action, organism's *effect world* (*Wirkwelt*).[28] The analysis of different living Umwelts enables us to understand their being in the world much better than the consideration of mere anatomy: by considering Umwelts, we are seeing life as subjective, and a germ of subjectivity is present in any life as its self. While it is most likely impossible to take the point of view of another being and experience the world as it does, constrained by the limitation and characterized by the senses vastly different from ours,[29] it is possible at least to assess the probable scope of the reach of its world and this way to learn more about it as a subject. The most celebrated example here is Uexküll's tick, whose world consists of a specific range of temperature, the smell of butyric acid, and the ability to borrow into a warm skin to suck blood.[30] The main point here is that it is the living being that defines its environment, at the most elementary level – by being genetically predisposed to register only a small subset of the elements present in it as stimuli and act upon them in a limited number of ways; the sophistication of the animal will define the scope of its environment.[31] Yet if we analyze this further, we would see that one can distinguish not only between individual Umwelts that, to borrow Uexküll's metaphor, follow living beings as bubbles imagined to surround them as they waver in the wind or fly around,[32] but also between types of environments that are significant for the evolution of freedom and characterize different types of living beings.

Any living being possesses an environment as described above. Yet many are passive in this environment. If a certain chemical compound chances to be in the vicinity of the part that is receptive to it, it will be absorbed by the organism. The organism, however, will not *perceive* the material at a distance, will not move toward it, will not *act* in the sense of changing its environment based on a drive that originates within the confines of its body. A tree falling in a forest and destroying a number of ants is very different from an anteater doing the same: the tree never sought to do it and thus never *acted* in any important sense. On the other hand, we can have a living being that is active in its environment: perceives some or even many of its elements, processes in its nervous system the data extracted by its senses, and actively seeks satisfaction to its urges by moving around and acting in the world. The world of such organisms is vastly different from the environment of passive living beings. It has perception marks (*Merkmale*) that are not immediately adjacent to the subject's body. It is a moveable feast, a bubble that the organism carries around as it moves. The distance between the stimulus and the response passes through some sort of processing that controls the organism both in terms of collecting data and acting upon it. Even the example of the tick, as simple as it is, demonstrates this point: the tick senses butyric acid at a distance, falls down, burrows through the mammal's skin – rather than remaining stationary and register only whatever touches it. In order to distinguish this type of environment from the one of the passive organisms, I will refer to it as a *world*.

We can have living entities that cannot consider themselves, the world, and themselves in the world: they either register their environment without being aware of it or are aware of the environment and themselves without being aware of the environment *as* environment, themselves *as* selves, and themselves in the environments as their own selves in their respective environments. An organism can feel hunger, or lust, or pain, and to act upon these feelings without being able to give any sort of account of what happened, to consider it *as* something that happened. It is this problem that made Heidegger deny the term world to non-conceptual beings, e.g., non-verbal animals, and preserve it for humans, as explicated in Part Two of *The Fundamental Concepts of Metaphysics*.[33]

Heidegger argues that only discursively intelligent beings can have a world that he defines as "*manifestness of beings as such as a whole*."[34] All living beings would have access to their environment, have an *Umwelt*, yet those who are capable of treating beings *as such*, consider beings to belong to the category of beings, or, in our terms, exercise metacognitive analysis, have a very different environment, Heidegger's *Welt*.

Heidegger's argument for this can be summarized as follows.[35] The naïve concept of the world is that of consisting of beings, things that exist. Thus, to speak meaningfully of the world, we need to speak of the accessibility of beings. Non-human animals have access to various elements of the world: they can feel temperature, see ant hills, smell rot, invoke memories of encounters with cats, etc. However, they do not have access to beings as such, they do not register beings as beings: fish cannot address temperature as the flow of phlogiston or as molecular movement, pangolins have no concept of an ant hill as they have no concepts at all, and dogs can say nothing of the nature of cats as belonging to the species *Felis catus*. Therefore, concludes Heidegger, non-human animals do not have a world.

This conclusion seems unwarranted because what he calls a naïve concept of the world is arbitrary. It is also naïve, as it denies a world to pre-verbal humans incapable of naming objects. In addition, it excludes from the human world those aspects that are not attended conceptually, either because of the lack of attention or the absence of proper categories. The process of riding a bike, as many other skills, cannot be adequately described verbally by most bike riders, if it can be described in such a way by anybody at all. Some exceptional experiences of the type vividly exemplified by Arjuna's encounter with the true appearance of Krishna in *Bhagavad Gita* cannot be described verbally; Mamardashvili referred to these indescribables as black holes "where whole peoples and large, vast regions of human life can be,"[36] invoking Kafka for the examples of situations that are "foreign to their own language and are not commensurable to humans."[37] However, it is hard to exclude riding a bike from the rider's world; the same difficulty applies to the experiences we cannot describe, such as revelatory experiences – while indescribable, they do impact us, at times rather strongly. Heidegger himself was probably deeply uncomfortable with his narrow definition of the world, which is evident from his treatment of animals as *poor in world* (*Weltarm*) – deprived of a world yet still not worldless.[38]

Heidegger's argument, though, does establish a different nature of the world of discursively intelligent being and the Umwelt of non-verbal ones. The Umwelt of non-verbal organisms, even those that are endowed with senses, movement, and self-control, cannot "refer beyond itself."[39] It is in this sense closed, locked within itself – its subject cannot consider it, only live within it. The human Umwelt, on the other hand, "is not only that which it contains – it is also open in the direction toward which it is not yet is."[40] Our Umwelt does not simply have an additional dimension – it can have any number of dimensions. As it considers itself, it can do so in many ways, and the number of such ways depends on the type of consideration, which has no limit in principle. Uexküll brings up an example of the forest and notes that there is no one forest – there is the forester's forest, the botanist's forest, the lumberjack's forest, and so on.[41] Yet all these are different from the tiger's forest in a crucially important respect: our forester can try and consider the forest from the botanist's standpoint, or at least understand that there is such a possibility, yet no tiger is capable of such a feat. Moreover: foresters and lumberjacks can create myths *about* the forest, invent new ways to conceptualize it, and thus change its meaning to those who buy into their myths. Yet it is not only the creation of new meanings that makes the environment of discursive organisms unlimited – the ability to consider the environment in different ways is the same ability that helps to examine these ways and look for the new ones. To emphasize the similarity of the Umwelt of non-verbal animals and discursive humans, Uexküll notes that just like the fly is oblivious to the spider's web due to its perceptual system, humans are oblivious to the cholera bacilli in water as ours cannot detect them.[42] Yet this is incorrect: humans are able to detect cholera bacilli in water by devising clever means to do so, which is itself a result of the ability to consider the hypothesis of the existence of cholera bacilli and the limitations of the means we have to detect them. The very fact that Uexküll writes about the cholera bacilli in a meaningful way demonstrates this point. It is this ability, language-enabled metacognition, that makes a discursive environment unlimited in principle. This warrants referring to it as a *universe*, something different from both the world and the environment.

When we analyze life as proceeding in its environment, we arrive at the concept of meaning: any object that enters into a relationship with a living subject "becomes a carrier of meaning."[43] Without changing its properties, water, a rock, a blade of grass, music, a flash of lightning become meaningful *for* a certain living being, whether conceived as meaningful or merely felt. This way, we can talk about life as interaction with the aspects of its surrounding that are meaningful for the living being on its own terms, or semiosis.[44]

Semiosis is usually defined as a process of creation, action, and interpretation of a sign.[45] The meaning assigned to the term 'sign' in semiotics is much wider than is common in philosophy: for example, it can refer to a molecule that for a cell indicates the presence of a nutrient:[46] the sign is a "relation whereby a receptive system orders its world."[47] Therefore, semiosis can be seen as a process that involves a living entity and any elements of its surrounding that carry some significance for it: wetness for plants, ultrasonic sounds for bats, rainbow colors and adverbs for humans, etc. To become a carrier of significance for an organism, a stimulus should fall within the physico-chemical range that the organism both registers and processes. The second condition is no less important than the first one: a dry leaf would not register a ray of light that a live leaf would, an animal whose attention is captured by a hot pursuit might not register a sound, and a word written in Hangul characters would be of zero significance for a person who cannot read Korean. This interaction is characteristic of every living being, from a single cell to a large hairy mammal. At the cellular level, these are specific molecules that can bind to the cell's receptors that are carriers of significance;[48] for a mammal these are the elements in the air it breathes, the smells it processes, the warmth of the body of other animals with whom it cuddles, the vocal warning signs, etc. For discursive animals, the semiosphere expands to include the layer of thinking, the *noosphere* of human knowledge and wisdom that has a seemingly unlimited scope and is generated by the living thinking entities themselves.[49]

Organisms' Umwelts can be interconnected: a number of organisms can share larger or smaller segments of their environments. Flowers and bees register sunlight, wolves and sheep can smell, dogs and humans can feel affection. It is the intersection of the Umwelts that allows for any sort of interaction between living beings: if their environments do not intersect, they cannot interact in any way, be it predation, symbiosis, or mating. This way we can talk about a *semiosphere*, the sum total of all interconnected Umwelts.[50]

Organism's survival depends on its interaction with the world: as an energy-consuming entity living in the world of other energy-consuming entities, each organism must use nutrients and can itself become a nutrient for other organisms. This makes an organism's relationship to its Umwelt a subject for evolutionary consideration. This also makes semiosis, the sum of organism-world relationships, relevant to the question of the directionality of evolution and the emergence of choice and freedom.

I suggested earlier that evolution moves toward greater complexity, defined as the number of part and the relations between them,[51] or morphological complexity. This would inevitably relate to what Hoffmeyer calls *semiotic complexity* that reflects "the depth of meaning communicated or interpreted by living systems"[52] – the capacity to interpret more complex indicators and a larger number of indicators. The more complex an animal is, the larger is the number of stimuli in its environment that have significance for it: with increasing complexity comes a larger, richer Umwelt. In this case richness and size are one and the same thing: when an animal acquires senses, it gains access to additional dimensions of stimuli, and thus vastly increases its world in comparison to the one of plants, for example.

The size of organism's world is positively correlated with the scope if its choice. The defining direction at first is that from the organism's morphological complexity toward its semiotic complexity: little interaction with the environment results from the limited number of ways for the living entity to interact with it. However, with the development of senses, the size of the organism's world starts playing a more serious role in its survival: access to a larger array of stimuli gives the organism more ways it can maintain itself in the world. Acquiring motility and sensation would be a good example here: an animal, differently from plants, can move toward food and escape danger – its Umwelt includes a variety of options related to these activities. Animals who have the means to develop a richer Umwelt and demonstrate better skills in interacting with it, or higher semiotic competence,[53] would do better in terms of detecting and accessing nutrients, registering and escaping danger, discovering and approaching mates, etc., tasks that are directly related to survival. The higher we climb in terms of complexity, the richer is the Umwelt, and the more significant is the role of the semiotic capacity;[54] naturally, the role of physiological structures that enable such capacity becomes greater as well – thus the development of the nervous system in higher animals. Once we get to humans, semiotic capacity seems to dwarf morphological complexity as far as their significance for survival is concerned, as I will argue later.

One can distinguish between individual Umwelts that belong to various living organisms and some sort of an objective space, what von Uexküll referred to as *Umgebung*. Following his line of thinking, though, one might argue, as Agamben does, that there is no such thing as an objective space: it is merely the human Umwelt that we are referring to by this term.[55] There seems to be, however, a significant difference between the human consideration of the world and other living beings' Umwelts. The objective space, as constructed by the human mind in its attempts to explore the world as it is, is characterized by striving to make it independent from the contingencies of human cognition. Von Uexküll, as a devoted Kantian, perhaps more devoted than Kant himself ever cared to be, would have objected to this suggestion. However, what is important for our purposes is the very attempt to develop an objective, subject-independent account – something that can be achieved only with language, through the ability to go meta on every mental content. This will be further analyzed later in this chapter.[56]

Once the capacity for choice appears, the significance of the Umwelt becomes much higher: the more alternatives there are, the more meaningful is the choice. Whatever increases the Umwelt and thus the scope of choice becomes even more advantageous evolutionarily. Organisms with an Umwelt that is significantly larger than that of other living beings will become the champions of evolution: they will have more options whereby to survive and reproduce. In terms of freedom, the capacity to define new alternatives becomes an evolutionary advantage. Some new alternatives can be defined by applying generalizations of existing images to available sensations and considering the results in light of existing interests. Putting one box on top of another in order to reach for a banana that is placed high up can serve as an example: the mental operation behind this exercise requires no abstract conceptions, only generalizations over sensory images. However, alternative ways of looking at the relations between things cannot be achieved by generalization over sensory images. To think of cultivating bananas, one needs a conception of plant growth that pertains to the way fruits come about; it is this sort of thinking that will leads to experimenting with the banana tree roots and using segments of its rhizome to grow new plants. This is achievable only with discursive metacognition.

Care

Living organisms are always interacting with their biosphere to sustain themselves, to the very least – absorbing nutrients and releasing waste. In other words, all life is characterized by "selective sensitivity" that is an "instrument of self-concern."[57] Therefore, all organisms can be said to have an interest in their environment being such that it caters to the continuation of their lives, whether this interest is felt by the organism or even conceptualized by it as an interest or not. Life, thus, can be characterized by the attitude of interestedness – this follows from its needy nature and the structures it has that are necessary to satisfy its needs. The way the needs are satisfied would be the way the attitude of interestedness is carried out.

Heidegger develops the conception of this disposition from observing the basic attitude of human Dasein, the way of our being in the world, to be anxiety (*Angst*).[58] Anxiety as an attitude is not focused on something specific yet exists *a priori*, is a necessary conditions for our interaction with the surrounding world, the world into which we have been thrown (*geworfen*; thence the condition of thrownness, *Geworfenheit*). For a human self, discursively self-aware, it is a feeling that discloses the distinctness of the self from the world to the self through experiencing "a collapse of familiarity, significance, meaning, mattering, or the world," per more popular interpretations of Heidegger.[59] Yet if we are to take an evolutionary view, we might consider anxiety in non-verbal yet nevertheless experiencing animals as a mere feeling of a rift between the self and the world, a feeling of the gap that exists between what the self needs and what the self has, the needs and their fulfillment – the feeling of dissatisfaction, of lack. The ontological structure of the way of existence for which anxiety is the basic attitude is care: concern (*Besorgnis*) for what is going on in the world. This makes perfect evolutionary sense: a needy entity that should act in order to survive, e.g., get food, would not survive if it is not geared toward being concerned for the surroundings it needs to survive.[60] All living beings owe their lives to interacting with the environment in which they are placed – in this sense the conditions of thrownness is characteristic of all life, whether it is aware or unaware. Therefore, if we want to understand the attitude of care, we need to develop it from interestedness rather than human anxiety which was Heidegger's sole concern.

If we take interestedness as our starting point, it will lead us to the necessity for every living being to be disposed toward fulfilling its needs. Such disposition can be passive and implemented through structures that selectively absorb materials that happen to be in the organism's immediate environment. It also can be active, where the organism possesses an attitude that propels it toward seeking satisfaction to its needs. It seems that the term 'care,' as signifying a somewhat active attitude, is more proper for the latter kind of life forms. Finally, in self-aware organisms a meta-attitude of care is possible, where the organism considers the merits of different ways of interacting with the world, caring for satisfying its needs and, perhaps, for other things, e.g., for doing what they see as right, proper, evaluatively preferred – discursive organisms can think about states of affairs and relate them to criteria of their own making. These three ways of being disposed toward the satisfaction of needs in the word will be later described as characteristic of unicellular organisms and plants, non-verbal animals, and humans.

As for freedom, Heidegger ties it to anxiety.[61] Since anxiety is a general attitude that is not focused on specific aspects or elements of the world, it provides a horizon of choice, in Heidegger's words – "discloses Dasein as Being-possible."[62] From here, if we are to analyze care as the ontological structure of interestedness, we would see that the form of freedom would be tied to the form of care. Living entities that are not characterized by the attitude of care cannot be free. Those that care can be, and an agent that can consider its own attitude would be characterized as "[b]eing-free for [*Freisein für*] the freedom of choosing itself and taking hold of itself."[63]

Plasticity

Plasticity is the potential for differences in behavior, morphology, and physiology between individuals having a similar enough genotype to be classified as members of the same species, where the particular behavior, morphological change, or physiological variant are not due to the differences in the genotype.[64] The genotype here, instead of giving rise to a specific characteristic, is enabling the organism to develop and demonstrate this or that trait in response to the environmental demands or for another reason. The most striking example of morphological and physiological plasticity is, perhaps, the Mexican axolotl (*Ambystoma mexicanum*). This salamander does not complete its metamorphosis into a lunged adult as long as the environmental conditions in its lake are suitable, and it can live its entire life as a fish. Yet, if the lake dries up or becomes otherwise detrimental to its thriving, the axolotl develops lungs, puts its limbs to pedestrian use, and walks to another lake.[65] Biochemically, this ability relies on the stem cells that can be activated in ways that give rise to their mitosis as different cell types. Axolotls possess a large amount of such cells, and this makes them capable of other feats, e.g., regenerating lost limbs.

Plasticity, though, is not limited to exotic cases. The more we go up the Aristotle's ladder of life, the more plasticity we find. Unicellular organisms are capable of little change, limited mostly to the amount of mitochondria. However, plants and fungi demonstrate plasticity in terms of their shape: they can grow larger or remain smaller, spring more or fewer branches, remain stand-alone or form wide networks of nutrient exchange,[66] have a root system that spreads around or goes deeper, depending on the availability of nutrients. In animals, the plasticity is less morphological but more neurophysiological and behavioral: the number of limbs usually does not differ, yet certain aspects of the nervous system differ quite a bit as a result of divergent experiences of individuals, a phenomenon referred to as *neuroplasticity*.[67] In more developed animals, including humans, behavioral plasticity is omnipresent: the ability to learn, procedural in non-verbal and animals, procedural and declarative[68] in humans, causes animals to behave differently following certain interactions with their environment; in humans, such behavioral changes can result from analyzing such interactions and starting to consider them good or bad, necessitating this or that response.

Plasticity can be seen as a serious evolutionary advantage. Variable environments favor incomplete, i.e., plastic design, as Dennett noted.[69] A phenotype that is responsive to environmental conditions and alters the traits relevant to survival accordingly is likely to lead to much higher survival and thus reproduction rates, so it will pass forward the genotype that gave rise to it. For this reason plasticity, once it appears, becomes a factor that changes the structure of evolutionary pressures: now possessing a more plastic life form becomes more advantageous than having a less plastic one. With less plasticity, changes in the environment might mean extinction where more plastic phenotypes have a chance. Thus, the degree of plasticity becomes an evolutionary issue, and this makes it necessary to consider types of plasticity.

One can categorize plasticity into two broad types: genetically determined and choice-driven. All the examples of plasticity in plants and animals that are incapable of choice are of the first type; this type of plasticity is also preserved in the animals that can choose between alternatives, alongside the choice-driven one. Having a genotype that enables growing lungs, changing coloration, or having a root system common with other aspen trees in the grove is something that is defined in the individual's genotype, and when the right environmental trigger presents itself, the response follows: proper proteins are generated, stem cells give rise to specific cells, roots spring out, etc. Similar processes can be observed in neuroplastic changes in mammals: when damaged organs stop sending neural impulses to the brain, those areas of the cortex that normally respond to them shrink, and the areas receptive to inputs from other organs expand on their behalf, making these non-damaged organs more helpful – a phenomenon deemed compensatory plasticity.[70] Similar effects can be seen upon training, both in the nervous system and in muscles. However, when the capacity for choice appears, plasticity acquires additional meaning.

Plasticity is a necessary pre-condition for choice. Without an organism's ability to change, e.g., to behave in different ways, the capacity to choose between alternatives would be meaningless. It also enables the choices made to have lasting effects. Without muscles being susceptible to growth as a result of weight lifting, weight lifting as an activity would never take off. Without the ability of the brain structures, e.g., synaptic connections, to change in order to store images of what the organism was exposed to, there would be no memory and therefore learning. However, in an organism that can choose between different behavioral options, morphological and physiological plasticity, and genetically determined plasticity in general, starts playing a lesser role in survival. It might be very important for a plant to be capable of surviving when rains do not come, e.g., by sprouting longer roots. Yet for a goat the ability to find patches of grass and move toward them, which requires choosing the right path, is more important than, say, the ability to abstain from food for longer stretches of time – food never comes his way, he has to go and find it. This would first lead to the emphasis on neuroplasticity, the enabler of the ability to remember, retrieve, and associate images.

Yet at the same time and for the same reason explained above the capacity to choose between alternatives will become more important evolutionarily, and the neurophysiological structures that enable it – a better trait to possess in terms of survival. The genotype that is favored here is the one that enables choice rather that specific behavioral patterns; this is the stage where instincts fall out of favor with evolution, as will be described later.

With language, behavioral plasticity gives way, even though not entirely, to what can be called mental plasticity. At this stage, the individuals have the most powerful means to adapt to change – they can try and understand the changes and act according to their understanding: migrate to warmer climes, build shelters, produce antibiotics, modify their environment and even their own genes. Non-discursive choice undoubtedly contributes to survival. By being capable to choose between different paths toward food, select this rather that that material to build a nest, choose to use a stick for this rather than that purpose, or alter the relations between the members of the clan an animal can adjust better to the changing environment within its lifetime, without waiting for the natural selection to leave only those individuals that have the right instinct for the current environment. The alternatives such choice considers may come as a result of being exposed to co-occurring stimuli, developing and remembering cognitive maps, social learning in more cognitively advanced animals, figurative imagination, or from other sources that are accessible to non-verbal cognition. With discursive metacognition, the ability to consider the relations between the own self and the world leads to the development of radically new ways of coping with the environmental challenges and thus opens a new horizon of adaptability. Now agents can think about the nature of the connections between various phenomena, not merely remember the phenomena they were exposed to and the associating them. For example, connections between things and events can be analyzed in terms of causal laws, which lead to the attempts to manipulate the cause to facilitate the desired outcome. This way, humans can think of the causes of feeling cold and possible ways to alleviate the associated discomfort, thus designing clothing; think of meals in terms of their quality, and then try and improve it by changing the diet and the ways they prepare food; consider different possible causes for diseases and devise cures; think about their interactions in terms of the organization of society and try new ways to organize it; and more. The adaptability here does not merely expand its reach but makes adaptations potentially unlimited: now the ways of coping with the pressures of natural selection can be altered and new such ways can be developed.[71] As a result, the structure of the evolutionary pressures themselves changes, as it will be described later in the section *The Distinct Nature of Human Evolution*.

Discursive choice, or freedom, is the most important plasticity of humans; and this discursive choice makes the neurophysiological structures that enable morphological and even some aspects of neural plasticity of lesser importance. Humans can make tools instead of growing larger limbs, print books instead of increasing the capacity to remember things, and create medicine instead of keeping a large amount of stem cells past initial development like axolotls. This way, animal choice and later human freedom impact the evolution of plasticity as much as they are enabled by it.

Variably Restricted Evolutionary Paths: Freedom in Design Space

The very character of evolution consists in the survival of those individuals that are more fit for their environment. These individuals pass their endowment to their offspring – the genetic endowment in the form of the DNA as well as some behavioral patterns and, in humans, knowledge through social learning. It is this mechanism that greatly restricts the available evolutionary paths by reducing the number of available alternatives: it is highly unlikely that some mutation or a recombination during meiosis will change a large number of genes massively, or that the cultural contents that pass down the line of generations in society will undergo too dramatic of a change for a large number of people on the planet. There are also constraints that are imposed by the biosphere, e.g., in terms of the availability of nutrients and other environmentally conditioned aspects of survival, e.g., the competition from other life forms.[72] These, however, are not the only factors restricting the scope of evolution: the more complex the organism is, the lower is the possibility that a massive genetic change will help its survival, too many traits are interdependent, and massive random changes in the genome are likely to upset the balance and make the individual not viable. And if the changes are not detrimental to survival yet are still not woven into the functional organismic whole, it is unlikely they will be preserved in the evolutionary future. If, for example, a certain plant, due to an extremely low probability mutation acquires a semblance of a sensory organ, it would be useless to it: the plant does not have a nervous system to process its outputs, nor does it have the organs to make good on the data it delivers, e.g., become better at feeding by moving toward food it now can sense. The same predicament applies to acquiring motility without senses and central control, as well as central control and motility without senses. The new trait will not help plant's survival while consuming energy, and will likely be lost in the coming generations.

Restricted evolutionary paths have been hitherto conceptualized within the model of the Tree of Life – "the graph that plots the time-line trajectories of all things that have ever lived on this planet."[73] The first drawing of such tree in the context of evolution comes from Darwin's *B Notebook*,[74] while a more advanced one appears in Chapter IV of the *Origin*.[75] Evolutionary trees depict the suggested lines of descent from assumed common ancestors, where each group of organisms belonging to this or that branch has only one presumed phylogenetic ancestor – another similar group. Ultimately the tree would come to a presumed first ancient organism.

Different ways were suggested to draw the evolutionary trees before the discovery of the DNA, e.g., based on morphology or main activity: producers like plants, consumers like animals that feed on plants, decomposers like bacteria that feed on dead organic matter, etc.[76] The problem with these is that they do not necessarily point to common descent: certain organs might be analogous in appearance and function but not homologous, i.e., having common descent; on the other hand, certain organs might not be analogous but still be homologous, like ape's hand, dog's paw, and whale's flipper. Phylogenetic trees that reflect common descent became possible with the analysis of the molecules characteristic of life, the structure of which is determined by the genes. This method, deemed *molecular phylogenetics*, was suggested by Francis Crick and perfected by Carl Woese.[77] Woese chose the ribosomal RNA (rRNA), a molecule that is present in all living beings, as his molecular clock, and calculated the degree of difference between different living beings based on the divergence of their rRNA components.[78] Based on these calculations, Woese suggested three domains of life: Bacteria, Archaea, and Eukarya, all having a common unicellular ancestor.[79] The space within which life can exist would be defined by the laws of physics – Dennett's *design space* for all material beings; yet the evolutionary paths along which the phylogenetic development can progress would be limited by additional factors – the laws of biology, including those of genetics, and the bio-historical ancestry of each individual organism – the branches of the tree.[80] This is why monkeys do not lay eggs – they split from egg-laying creatures so long ago that their genome does not code for the multiple proteins necessary for this complex activity; and if some appear suddenly in this or that ape due to an extremely rare and massive mutation, laying eggs will be such an outstanding activity in monkey's lifestyle as suggested by its other genes, that this trait probably will not be carried forward. This possibility is foreclosed for monkeys, to use Dennett's language.[81]

The Tree of Life, however, has been seriously shaken by the genetic research carried out in the last thirty years and now cannot be seen as an entirely adequately depiction of the history of life. Slow random mutations cannot explain the rapid change in the phenotype of bacteria that has been observed since the 1920s,[82] and bacteria that reproduce by simple cellular fission do not have the recombination of genes.[83] These observations were explained, however, after the discovery of the *horizontal gene transfer* (HGT) – conveying of genetic material from one organism to another not by the way of descent.[84]

There are several ways for genes to travel from one organism to another. Firstly, whole bacteria can be taken inside the cytoplasm of another unicellular organism, with its genetic baggage – a phenomenon referred to as *endosymbiosis*. This way, a bacterium could have entered an archaeon and become cell's mitochondria, while the ancestors of today's cyanobacteria could have lodged inside a unicellular organism that later gave rise to plant cells, with the descendants of the newcomer becoming chloroplasts. The support for this explanation relies on the molecular phylogenetic analysis that discovered significant differences in the RNA and rRNA of chloroplasts in algae, mitochondria in wheat, and more.[85] Secondly, there is endosymbiotic gene transfer: certain genetic materials can travel from the organelles to the DNA.[86] Thirdly, segments of the DNA can travel from one organism to another, e.g., from the microbiome, the microbial fauna in the guts of a mammal or an insect, to its host. The genetic material can be carried by viruses, get form one cell to another through cell fusion, and more.[87] In most cases such transfers would not bring any new traits, as they either transfer unused, "junk" DNA, or result in non-viable organisms. In some cases, however, they give rise to traits useful for survival, e.g., antibiotic resistance in bacteria.

Some scientists today see the horizontal gene transfer as invalidating the concept of the tree of life: the different branches do not appear to be so separate any more, and thus Dennett's paths in the evolutionary design space become intertwined. This precludes the history of life from being properly represented as a tree,[88] calling forward the metaphor of the web of life. However, this does not seem to deliver a mortal blow to design paths.

While horizontal gene transfer has been observed in many organisms, from bacteria to humans, it is most easy among bacteria and archaea.[89] In fact, it is so pervasive at this level that some researchers suggested considering all bacteria on Earth as one large species with a common gene pool.[90] Bacteria have a relatively short genome that defines their way of life which is also quite simple in comparison to plants or animals. Bacteria do not learn in any shape or form, their functioning is fully prescribed by their genes as responding to the chemicals in their immediate environment. Their survival relies solely on what their genome has to offer. For them, HGT is the best way to produce multiple variations to be tested by natural selection – this is their "font of possibilities."[91] Yet the longer the genome is, the harder it is to make a significant change in the phenotype by adding a small segment to it: long genomes of more complex organisms produce a variety of proteins that give rise to structures interacting in complex ways. Making a change here by simply adding a gene sequence is more likely to break the interaction of parts and processes, thus rendering the organism non-viable. Thus, the higher we climb up the tree of life, the later we proceed with the evolution of life, the lesser would be the impact of the HGT – not the one that took place earlier in the history of life but the one happening on the daily basis. In fact, advanced organism have special ways to counter microscopic invaders, including those that can potentially add segments of their DNA to the human one – the immune system.

At a certain evolutionary stage the role played by the invading genes seem to be taken over by other phenomena. The evolutionary advantage of the lateral transfer of genes is clear: it consists in rapid acquisition of traits. With conditioning, an alternative and physiologically more accessible way to do the same becomes possible: there is no need for complex interactions of the protein-producing cellular machinery, with all possible incompatibilities, for a house cat to learn that fighting opossums is a bad idea – one encounter might be enough. An animal that can be conditioned is more adaptable: it has less of an inclination to repeat the same mistake twice while it is inclined to repeat successes. At this point, what becomes useful passing on are the genes that support conditioning, rather than the genes that underlie functional traits: the genes that support nervous system and protect its physiological structures. Now extra genes entering the organismic structures supporting inheritance become worth protecting against, as the organism can acquire adaptations in other ways. Yet animal learning is not confined to conditioning, it becomes more sophisticated with further development of the ability to manipulate retained images and generalizations over them. The very same neurophysiological structures that support conditioning, i.e., the elements of central nervous system that make possible generalizing over retained sensory images and linking them, develops further and gives rise to cognitive maps and other similarly advanced capacities, including learning from others: procedural learning, where the organism does not yet understand that it is learning from others or learning at all, as it has no discursive metacognition. However, this animal's ability to acquire new skills goes beyond mere conditioning, remembering that the stimuli S_1 and S_2 are related to feeling F and invoking F when intuiting S_1 and S_2. Such an animal that can also manipulate generalizations of images, recombine them, and relate the result to desire, would be more adaptable than an animal that needs to experience S_1 and S_2 together with or prior to experiencing F. An animal that can observe its conspecifics and relate their behavior to its outcomes, and then imagine itself behaving in a similar way and relate it to the observed outcome, would be capable of social learning: acquiring behavior from other species members. At this stage, behaviors can transfer from one generation to another by social learning, as it is common among many mammals. Social learning here is always horizontal: the behavioral patterns are not inherited biologically, it is the ability to learn them that is genetically heritable.

At this stage, the evolutionary paths are much less restricted than in living beings not capable of conditioning. Whatever the animal learns throughout its lifetime can greatly impact the factors of its survival. The most common example would sparrows and cats. The ancestors of both learned that co-dwelling with humans can provide for food and shelter, and both now would have serious trouble surviving without *homo sapiens*. Avoiding predators becomes less important for sparrows, yet increased dietary span becomes more important. As for cats, docility tramples all what other felines need the most, e.g., keen senses and speed; only in an environment where they have readily available food and minimum natural enemies can cat escapees survive. Similar examples of changes made by ancestors that impact the variables of the survival equation abound throughout the animal kingdom, from Darwin's finches to marine mammals. However, genetically determined physiological constraints still play a major role: once in the water, whales' path back to land would be highly problematic; same applies to sparrow's ability to become sea dwellers. Behavior is still dominated by the genetic endowment, with the opposite influence being relatively weak.

With pre-linguistic choice, the influence of life style on behavior becomes more pronounced. Now the repertoire of behaviors greatly increases, and the effect just described becomes stronger. Conditioning, even backed up by cognitive maps, is passive: an animal's mind connects between what happens to it. With the advent of choice, individual animals will make different choices leading to varying degrees of success, thus providing an equivalent of different variations. Those who make the best choices have a higher change to pass their genes on. Those who are capable to learn from their mistakes have the best chance. In a sense, individual animals with a wider repertoire of choice and the ability to learn from their mistakes in terms of the survival advantage are equivalent here to microbial populations that are more susceptible to the horizontal gene transfer and quickly produce multiple variations, like the pneumococcus that can rapidly absorb antibiotic-resistance genes or become more virulent due to HGT. These are the individuals who pass their genes on, providing for a chance of some individuals in the next generation to become more cognitively apt. Yet still the limits imposed by genetics are of a principal significance: individual non-verbal animals cannot consider their successes and mistakes, advantages and limitations, analyze them, and try and chart the best path forward based on this analysis. In them, for example, it makes perfect sense to assess cognitive capacity: the animal who is capable of storing more images, of associating them in more ways, of being able to retrieve and apply these associations to sensed images, would be more successful. In this sense, typical cognitive tests are applicable to non-verbal animals because the aptness of their response is directly related to their prowess in terms of the memory capacity and the processing of sensory data.

With discursive metacognition making its evolutionary appearance, the context changes dramatically. Now we can generate knowledge, cognition about other cognitions that is not limited to sensory contents: analyze our experiences, try and understand the connections between events, and organize our behavior accordingly. This knowledge becomes easily transferrable with language: there is no need any more to experience everything yourself or to try and learn it in a hard way, through association, from other species members. Instead, one can ask or read a book. Here we also acquire human, discursive freedom, the ability to make choices regarding ways of behavior rather than merely pertaining to specific actions, have access to many more alternatives and create new ones. The ways of acting here have an aspect of universality, pertaining to principles and guidelines that do not so much dictate specific behaviors and the sensory characteristics of situations where they will be carried out as indicate frameworks for making decisions that can apply to an unlimited number of situations. Traditions asking for oracular guidance and appeasing gods in cases of difficulty, paradigms of looking at the world as organized hierarchically, the Golden Rule, and the empirical approach to learning about the world are some examples of such ways. Each restricts the way we make decisions: an individual or a society that is steeped in the hierarchical paradigm is not likely to abandon it without some serious struggle. However, discursive intelligence enables us to describe each of these ways so that it become comparable to others, judge them according to the criteria we choose, and abandon one way of behaving for another if we decide to do so.

The range of choice becomes so much greater than that accessible to nonverbal animals that genetic constraints lose much of their restrictive power. With the development of culture, science, and technology, constraints of physical strength and distance become of lesser significance. Physical health becomes a field of conscious choice, thus rendering the impact of genes that determine our physique progressively less significant: what has been a death sentence in a proto-human stage, e.g., near-sightedness, becomes not even a cosmetic issue with laser eye surgery. Memory and associative capacities also lose much of their crucial meaning: a person who knows how to store externally and find the right information is much more fit for the world of the twenty-first century than one who can remember many things. This, because the amount of information available to the former greatly exceeds the one available to the latter. The sheer processing capacity of the brain becomes less important, giving way, perhaps, to the number of levels of meta that the person is capable of, to one's metacognitive ability, the ability to think about pieces of knowledge, methods of acquiring knowledge, about the principles of behavior and the principles of the thinking involved in it. This puts a question mark over the usefulness of the standard cognitive testing, including IQ tests, for humans. Yet this also points to the profound change in the way of thinking and decision making that is brought about by discursive metacognition. The amount of information that exceeds the retaining capacities of human brain necessitated the development of clay tablets and thumb drives. Yet it is the ability to deal with universals, to think about the novel ways of solving problems, that made them possible: developing means to retain information requires comprehending the universal category of information, a category that is not limited to a certain image or a specific set of images. Only when we can think about information as a category can we start looking for a solution that will satisfy the criterion of being able to preserve any specific piece of it, examine various ways of registering it in light of this criterion, be it pictograms on cave walls, characters signifying words preserved on clay tablets, or magnetic records. This ability also created the needs to deal with so much information: chimps do not need Hammurabi's laws because they cannot grasp the notion of law, and goats have no use for accounting programs because of the same reason.[92]

With human freedom, we are able to define the evolutionary paths ourselves, according to our own views of how we should progress: we can think about ways of doing things, examine them against abstract criteria that are universal and range over all possible ways of behaving, and consider these criteria themselves, which requires language-enabled metacognition.[93] This does not mean that we can, at a certain moment in time, foresee the consequences, not even the end of the road but the next turn. Yet we can decide which road to follow. The whole of Dennett's design space, delineated by the laws of physics, is open for humans, and history restricts us only to a degree we have not yet acquired knowledge of how to overcome it.[94]

FROM UNICELLULAR TO MULTICELLULAR ORGANISMS

Unicellular organisms can be individuated by their biochemical unity: the processes that characterize the unicellulars are distinct from those that are present in their environment.[95] The unity of these processes also leads to spatial unity: single cells, be they prokaryotic bacteria or eukaryotic protists, are separated from their environment by a semi-permeable membrane. This is the bare minimum to answer the specification of life: unicellulars lack parts, the organelles of eukaryotes, e.g., mitochondria, are quasi-independent and resemble endosymbiotic beings from which they likely descended, and follow their own agenda. The only need the unicellulars demonstrate is nutrition: they are too simple to ask for more. Therefore, their Umwelt is no more than an environment. Unicellular organisms have no perception world, as they do not perceive. They also have no effect world, since they do not act: they impact their environment yet do not carry intentional action, lacking the structures necessary to form an intention. The connection the unicellulars have with the world is that of interest yet not care: they cannot care as they lack sensation and action; the interest they have, nutrition, can be imputed to them yet is by no means felt or otherwise experienced, since they have no sensation. Accordingly, everything in the unicellular's structure and way of functioning is determined by its genotype, whether inherited from the mother cell or altered by the horizontal gene transfer.

Unicellular organism's only lack, then, is access to nutrition. Single cells are small, they have no capacity to store much nutrients for future consumption. They also lack the capacity to move toward nutrition without, say, being carried by water – yet they need to be in an immediate contact with nutrients for the latter to be absorbed by the cell's semi-permeable membrane and become useful for the cell. This makes survival problematic when the environment changes and the nutrients stop being continuously available. One coping mechanism that the unicellulars possess is quick multiplication through mitosis: when the nutrients are available, the cells multiply. This efficient multiplication that does not require excessive resources is the major species survival mechanisms for simple life. However, it has a serious limitation: the more unicellular organisms multiply, the faster they consume the available nutrients – thus undermining their own survival.

Quick multiplication, though, has a significant evolutionary advantage: it produces a large variety of phenotypes, the living hypotheses that offer themselves to the test by natural selection. Two factors greatly increase the genotypical and phenotypical variability of unicellular organisms beyond quickly multiplying mutations: horizontal gene transfer and organismic simplicity. Horizontal gene transfer has been addressed earlier. The way simplicity contributes to the variety is more subtle.

Unicellular organisms have little complexity. They have significantly fewer parts and dependencies between those parts than multicellular organisms. This opens a wide horizon for phylogenetic change: the same quantity of genetic alteration, i.e., the same number of genes changed, will have a much greater impact on an organism that has fewer functions, since fewer genes are required to make a change for a simple organism without making in non-viable. In a more complex being, the same amount of changes has a much higher chance to corrupt crucial life processes as complex functions rely on multiple genetic sequences: they depend on many other bodily processes.

The great many phenotypes the multiplication of unicellular organisms produces demonstrate a variety of ways to adapt to the challenges they face, many of which are about being parasitic on other organisms. Yet, given the large spectrum of adaptations and the pressure toward complexity that has been discussed earlier, some would be evolving toward greater complexity, making the appearance of multicellular organisms highly likely.

Multicellular organisms are thought to have evolved independently thrice: as animals, plants, and fungi.[96] Yet what we are interested in here are the paradigmatic options, the development of the different types of life forms that reflect evolutionary strategies as relevant to the emergence of freedom. Slime mold can serve as an example of a junction where such different strategies diverge.[97]

When the bacteria found on the forest floor, slime mold's nutrition, are plentiful, its cells behave like independent unicellular organisms. Yet when confronted with food scarcity, individual slime mold cells issue a special chemical compound, cyclic adenosine monophosphate; it is found in human cells as well. Other slime mold cells register its presence and respond together, a phenomenon called quorum sensing: they move toward the signal, in a fashion that resembles that of plants that follow the light – hard to suspect slime mold of any conscious data processing characteristic of animals endowed with perception and a nervous system. Once they congregate together, though, they start behaving like a multicellular organism, advancing in unison in a formation referred to as a slug. If that does not help them in getting food, more chemical signals are issued, and slime mold cells start differentiating: some form stalks and die in the process; other create fruiting bodies, or sporangia, where haploid spores are produced by meiosis and can be disseminated later by wind and water much like in fungi.

This amazing behavioral and morphological plasticity of the slime mold is of special importance. Slime mold belong to the kingdom of Protista, a catch-all category where single and multi-cellular eukaryotes that are not fungi, land plants, or animals are placed. Protists have very few common characteristics[98] and many of them are genetically related more closely to various fungi, plants, and animals than to other protists.[99] All this leads us to the conclusion that the genetic code that underlies possible adaptive strategies in protists later plays a role in both plants and animals. This includes the amazing behavioral plasticity of the slime mold, its ability to morph into multicellular structures and to differentiate its initially uniform cells.

Following the analysis of slime mold's behavior, the move from unicellular to multicellular life can be explained by the chemical communication between cells, leading to them working together as one organism and differentiating their roles. Conceptually, we see how lack – of nutrition, in our case – drives both the diversity of adaptations and greater structural and behavioral complexity. And with greater complexity comes the potential for a more sophisticated self, one that can evolve into an entity that is aware of available alternatives and can choose between them.

The multicellular life form that emerged first from the unicellular life can be seen as *bare life* – a life "separated and excluded form itself."[100] It has the minimal necessary specification of life, enough only for determining that the entity is indeed alive rather than inanimate. This is a life that does not have awareness of its surroundings or of itself. Yet this sort of life has the potentialities that can develop into the different life forms. For example, slime mold in its multicellular instantiation demonstrates some rudimentary sort of movement that can later develop into animal motility, as well as stationary production of spores that can be carried by wind and water – something that is characteristic of the plant life form.[101]

Dennett suggests categorizing survival strategies into two broad types, what he calls *Maginot line* and *Guerilla* warfare.[102] The first strategy is the conservative one, that of the minimal possible increase in phenotypical and behavioral sophistication. This is the strategy of the least taxing adaptations: slowing down the metabolism instead of developing motility as a way to cope with the lack of nutrients, developing armor instead of faster movement, etc. The second strategy is that of increasing sophistication, acquiring complex traits: motility, senses, cognitive control over the organism. Assuming more or less random variation of the genotype in generation n+1 within the limits posed by the one of generation n, or the design paths available for the immediate descendants of generation n, we can surmise that in generation n+1 one would find phenotypes that are both of the conservative and of the sophisticated variety, relative to generation n[103] – both will be offered for the test of survival.

It is likely that the modifications of the genotype that are required for the conservative strategy are simpler than those that are needed for the one of increased sophistication. This, because complex would require genes that code for new structures which in turn might interfere with or depend upon other structures; same applies to the biochemical processes related to them. Moreover, with the increased sophistication we are likely to see higher demand for energy which will escalate the sophistication spiral.[104] Therefore, in terms of the evolutionary timeline we will likely see first the organisms that demonstrate conservative evolutionary strategy. Only later, after much more evolutionary trial and error, we will start seeing widespread examples of more sophisticated organisms.

The two strategies are exemplified by the plant and the animal life forms, where only the latter will lead to the emergence of freedom, as I will argue next.

THE EMERGENCE AND EVOLUTION OF PLANTS

Plants and fungi[105] are the outcome of the conservative evolutionary path, the collapse of the superposition of slime mold-like organisms that can demonstrate the features of both the conservative route and the path toward increased sophistication into the option of least change. This path is the easiest solution in evolutionary terms, as it requires much less genetic alteration that the other one. Thus, it would also be the first one to appear in terms of the evolutionary timeline. For a more significant change, sexual reproduction with its recombination of genes is a more likely vehicle, as it can produce more diverse genotypes quicker; for a smaller change, mutations accessible to asexual organisms would do. In the case of green plants, it was likely an instance of endosymbiosis that led to their appearance on the biological stage: cyanobacteria entrapped by another cell, giving rise to the chloroplast that enables photosynthesis.[106] Fungi kept even closer to the "plant mode" of the slime mold. Both plants and fungi cemented the ability to rely on absorbing the needed nutrients from their immediate environment through the semi-permeable membrane of their root cells and, in the case of plants, supplement it with absorbing the energy of the sun through their green surfaces.[107] This way, in terms of individuation, need, world, care, and plasticity plants show little change from what characterizes unicellular organisms, even though Earth's flora sports a great variety of species and is much older than its fauna. Same applies to lack: the only one plants have is that of nutrition.

Like the unicellular organisms, multicellular plants are individuated by the unity of the biochemical processes that constitute them. The way the plant develops is defined by its genotype, which, due to the more limited impact of horizontal gene transfer than the one we see in microbes, is more uniform across individual plants. It is the genotype that defines the different materials that constitute the parts of the plant and their general relation to each other – as any organism, plants are self-organizing entities, their development follows the steps encoded by their own matter.[108] Here, however, the environmental influence's role in shaping the plant is more significant than in the unicellular domain: for most plants, the number of parts, e.g., branches and roots, and their orientation in space is determined externally by the availability of nutrients, the position of the sun, etc. In this sense, plants are less definite than bacteria and unicellular eukaryotes – differently from them, plants have an indefinite form.[109] The use of the term *form* here goes beyond mere shape – plants' way of becoming is rhizomatic.[110] While it follows a certain principle encoded in the plant's genome, it is uncontrolled in its proliferation. Plant's parts are connected by chemical compounds released by each of them, as it is the case at the micro-scale with the unicellulars, yet the plant organism is not a unified whole acting together – it lacks functional unity. This is the reason that makes it so hard to distinguish between genets, genetically identical units of plant life, and ramets, plant individuals: in cases like moss or interconnected root systems, there is no way to point out individual plants, as there is no unit acting as one entity. Therefore, as any other rhizomatic system, e.g., the Internet, the plant has fragmentation rather than unity as its individuating principle.[111]

This lack of unity is crucially important to understanding the plant life form, as it precludes plants from any sort of action: the latter requires the organism to be functionally unified, to control itself centrally, to act as one entity.

The proponents of plant intelligence who credit plants with purposeful behavior, future-oriented movement, competitive action, and even choice, argue that plants are individuals in a way similar to that of animals: unified systems capable of acting as one entity.[112] This is a crucial point for them, as if plants are not functionally unified individuals, they cannot act as such, or act as individuals – cannot act at all. The argument advanced in support of plants being unified individuals relies on the different parts of the plant communicating with each other through chemical messaging similar in terms of the compounds involved to the one existing in animals. This communication leads to changes in the biochemical processes in various parts of the plant. For example, botanists observed that changing conditions, e.g., removal or roots and leaves, lack of water or minerals, etc., "give rise to specific changes in growth and development elsewhere in the plant."[113] This, however, does not seem to be meaningfully different from the quorum sensing in bacteria – yet it does seem very different from the way any animal organism controls itself, where one processing center sends commands to different organs that all work in unison, which is clearly demonstrated by animal movement. Plants lack the structures for such self-control and indeed do not demonstrate it.[114]

This lack of self-control structures can be explained evolutionarily. The mode of plants' relationship with their environment is that of unmediated metabolism: plants and fungi absorb nutrients from their immediate surrounding, mostly inorganic.[115] This lack of distance between the organism and its nutrition would create little evolutionary pressure to develop senses: there is no need to perceive anything at a distance as the nutrients come into contact with the absorbing cells. The absence of senses and the immediacy of nutrition are unlikely to lead to the development of any active stance toward the world, of any structures that enable action, e.g., movement toward the perceived source of nutrition. Both sensation and movement are resource-intensive, as they require energy. Moreover, sensing and moving require central control structures to make sense, i.e., to have movement directed toward the sensed stimulus and corrected based on the sensed changes in its position; this requires energy too. Mutations that point in this direction are unlikely to survive in an organism that is equipped with a way of absorbing its nutrients from its immediate environs. Instead, plants survive by evolving structures of intensifying or slowing down growth, slower metabolism, and producing copious amount of seeds which makes reproduction more likely.

Some botanists suggest that plants do actually move, it is just that the scale is much slower than the one with which we usually associate movement. Plants grow toward light, their roots spread in the direction of moisture and nutrients; Trewavas suggests that plants can navigate a maze by growing toward a source of light – a clear sign of intelligence.[116] This, however, is misguided: plants do not grow toward light, their branches do; same applies to roots. Since plants are not functionally unified individuals, they do not grow as a unit. As for navigating mazes, it is just that those branches that are exposed to more light grow faster, while the rest wilt. Similar considerations apply to the argument that plants can sense. While their parts, e.g., leaves and roots, certainly register environmental factors like the intensity of sunlight and the moisture of soil, they lack the "single field of sentience"[117] – a necessary pre-condition for sensation. Whatever different plant organs register might cause the release of certain chemical compounds that are received by other parts of the plant and trigger a set of biochemical processes, yet they lack central processing of whatever is registered by the parts which would warrant the term "sensation." Thus, plants do not have a world: neither Uexküll's perception world, as they do not perceive, nor his effect world, as they do not act. Rather, they have an environment, an Umwelt with which they interact, yet in which they do not act, are passive rather than active, like the unicellulars.[118]

Since plants are passive in their environment, we cannot meaningfully talk about care in their case. Plants' immediate metabolism precludes the development of any sort of attitude toward the environment. Such attitude cannot arise as plants are not functionally unified individuals – there would be no bearer for this attitude. Moreover, immediate metabolism leaves no evolutionary room for desire: there is no gap between the need and its fulfillment that would make the development of desire, with its high energy costs, evolutionarily beneficial for the organism, i.e., contributing to its survival.[119]

At first glance, plants are ultimately plastic: their very shape can differ dramatically following the availability of space and nutrients. This ability, however, is completely determined by their genome: plants cannot be said to learn in any meaningful way. The argument from plant learning is based on interpreting research data in a way that supports plant memory, defined minimalistically as the "ability [of the information stored as a result of prior experience] to interact with, and modify, the transduction pathways of new signals."[120] In support of this, Trewavas states that "the present state of development acts as memory for any individual plant because the same signal [environmental influence – MY] can have different effects determined when the plant, tissue, or cell receive it."[121] This, however, seems indefensible: developmental progression, determined by the genotype, cannot be seen as memory precisely because it is inherited, not acquired. Vernalization of seeds is another example put forward in support of plant memory and learning: exposing seeds to lower temperature causes changes in their flowering schedule and other parameters. However, this phenomenon does not amount to learning as well: the effect of vernalization is likely to be fully determined genetically, and in this regard it is similar in its effect to the developmental sequence. Leaning, behavioral plasticity, is linked to the capacity of the whole organism to benefit of the remembered experiences – which is impossible for an organism that lacks functional unity.[122]

Plants preserve and, in the sense of their capability to survive, perfects Agamben' *bare life*,[123] life detached from subjectivity, from any activity of the nervous system. Such life, lacking any physiological enablers of concrete, self-aware subjectivity, e.g., central nervous system whereby the organism can control itself, any activity that would be enabled by these structures and would necessitate them or at least make them useful, is highly unlikely to evolve them. Plants do not have the physiological structures to have access to alternatives: they have no senses or memory. They have no unified self that can act as one entity. With their immediate metabolism they lack desire, and thus – the evolutionary pressure to develop cognitive structures. Any mutations in the plant genotype that constitute a step toward sensation or memory are unlikely to make their bearers more fit for survival while consuming energy – and therefore are unlikely to pass on. Consequently, the development of any capacity for choice in plants is highly improbable. Plants, some 90% of the Earth's eukaryotic biomass, stand in their iridescent green glory as an example of an evolutionary path that is unlikely to lead to freedom.

THE EMERGENCE AND EVOLUTION OF NON-VERBAL ANIMALS

Plants, no matter to what specific species they belong, are quite uniform in terms of their individuation, need, world, care, and plasticity. All plants are individuated by their biochemistry and to a degree – by spatial unity; demonstrate the need for nutrition and nothing else; have an environment yet lack a world; do not show any attitude of care; and are plastic merely in terms of shape. On all these scales there is no evolution, phylogenetic change in the plant kingdom, despite the great variety of plants in the epochs past and today. There are no plants that are meaningfully individuated by something else, e.g., the history of their exposure to stimuli that impact the way they interact with their surroundings. No plant has emotional needs. Neither a humble moss nor a towering sequoia is active in the world in the sense of actively seeking nutrition rather than relying on what comes its way; consequently, neither can exhibit care. Finally, no plant is plastic except in terms of its external shape: no plant acts and has a certain behavioral pattern that would be unique to it. The animal kingdom, on the other hand, demonstrates significant diversity on all these scales, both in our days and throughout its evolutionary history. The latter also suggest the progress toward freedom, as I will try to establish.

Evolving Toward Greater Sophistication

The animal answer to lack is motility – evolving the ability to move toward the object of need, e.g., food. The collapse of the slime mold's superposition into the mobile way of life is what gives rise to the animal kingdom, thus starting the evolution toward ever greater sophistication. This is because motility requires functional unity that goes beyond the self-organizing unity of the unicellulars and plants – it necessitates self-governance. It is also the reason for the great variety of fauna in terms of individuation, needs, world, care, and plasticity.

Ismael, in her analysis of different types of systems, distinguishes between dynamical groups, self-organizing systems, and self-governing systems.[124] Dynamical groups are merely collections of parts bound by physical forces – mechanical, chemical, or electromagnetic. Compound molecules, rivers, mountains, plants, and galaxies can serve as examples here. Self-organizing systems are those where we can see the principle of organization embodied within the system components themselves, systems with a clear boundary between the inside and the outside. It is hard to think about any example of such system but living organisms. Microbes and plants certainly belong to this kind of system. However, in order to be motile, to move out of its own impetus as opposed to being moved by external forces, an organism needs central control, what Ismael deems "top-down regulation of behavior."[125] To move as one unit, the organism must have a way to coordinate all its parts. To achieve that, the organism starts using the structures and processes that are available from the earlier stages of the evolutionary development in a new way. The release of chemicals that are registered by other cells and trigger some activity on their part, e.g., quorum sensing in slime mold, becomes a system that makes the organism move as one unit – and this is the beginning of progressing from coordinating different unicellular entities toward having them as parts of one organism that communicate with each other chemically, are synchronized in a way that makes the organism working together, as one entity.

At this point, the individuation of an animal organism is no more merely biochemical. Firstly, its spatio-temporal unity becomes more pronounced than in plants. In animals we see no uncontrolled growth driven by external factors: Hegel's sun that is a plant's soul since it drives its growth, roots growing where nutrients and wet spots are available, the indefinite number of branches, etc. Since it needs to move in one unit, the animal's form would be more defined – with rare exceptions like the number of hairs, for example. Secondly, the animal becomes individuated by the unity of its functioning – at the very least, it moves as one unit. An animal's environment now acquires new meaning: since the solution of standing still and waiting for the nutrients to come by has been abandoned, the world becomes a source of information about nutrition to move to and satisfy the lack.

The structure of evolutionary pressures changes too. For an organism that moves toward its food, new characteristics become important. It needs much more energy to sustain its movement, so whatever it extracts from its immediate environment, i.e., oxygen, is not enough anymore. Evolving a more sophisticated solution for energy needs that would rely on extracting nutrients from the immediate environment does not seem feasible for moving organisms: they cannot develop a system of roots, for example, as it would not coexist with movement. Using energy stored in other organisms, i.e., evolving further the feeding structures that exist in many unicellulars that feed on microbes, seems to be an open route for evolution, and a promising one at that: it is helped by the presence of plants and other animals that can provide the hungry organism with nutrition. Yet this is not enough: movement, to be of any use for feeding, needs to be directed toward food. There must be an initial impetus to do so and a drive to sustain it until the need for food is satisfied; the source for both should be in the organism itself so it is coordinated with movement.[126] In order to move toward its prey, Uexküll's tick must not only be in need of nutrients, it has to be hungry, to feel something that would make it do the move: a tick that does not feel hungry would starve; the ticks that, by a genetic quirk, do not develop this feeling of hunger, or the drive to feed, go extinct quickly. Yet this would not be enough either: a hungry tick that has no way to sense in some way the presence of warm blood, e.g., by temperature and the molecules of butyric acid, will die as fast. Therefore, motility always goes hand-in-hand with sentience and emotion. These two, of course, are linked in a way the organism controls itself: there is no use for sentience that is not tied to organism's urge to eat and its structures that initiate and sustain movement toward the sensed source of nutrition.[127]

This is the first example of the escalating cycle of sophistication: the animal mode of life is such that its chances of survival from the very beginning of fauna's evolutionary history depend on structures that are more complex than those present in plants whose sessile existence does not require either sensation or urges, let alone self-control. All these are energy-consuming too – and this puts evolutionary pressure toward keener senses, stronger urges, and better self-control. Competition with other animals adds to that as well.

If we take stock at this point, we would see that the animal's structure of needs is different from that of plants: besides nutrition, the animal also has what can be called rudimentary psychological needs, e.g., hunger, that it seeks to satisfy. In fact, this is what the animal organisms is striving to allay rather than the need for food. Acting upon an unfelt need would require consideration of needs in the context of the organism, comprehending what a need is regardless of felt sensations or the lack thereof – something no non-verbal animal that lacks discursive metacognition is capable of. Instead, the animal feels discomfort and seeks to alleviate it. With sensation and motility, it acquires a world: no more an environment of a passive organism that stands, desireless, not seeking anything but relying on the presence of nutrients in its immediate surrounding, but an active quest for satisfaction of its urges through sensation and movement toward the sensed source of nutrition, coordinated by the centralized self-control. The urges pushing for moving in the perceived world are care – the first evolutionary appearance of what later develops in quite sophisticated ways.

These three characteristics – self-controlled motility, sentience, and emotion[128] – change the way the organism is individuated. In plants, we can talk only about formal subjectivity: we impute unity to plants, while they do not live their lives as unified entities that act as such. In animals, to the contrary, we have concrete subjectivity:[129] a "self which is for the self" – "shape idealized into its members," a unity of distinct moments pervaded by the said unity, a unitary soul (*Seele*).[130] The distinction between the self and the world here is something the animal is aware of, even if in a very rudimentary fashion: without this distinction, no goal-directed movement would be possible. With animals, we have "a more pronounced self [that] is set over against a more pronounced world"[131] – or a self against the world, a subject against the world of object.

The subjective self that evolves into existence with the tripartite emergence of self-controlled motility, sentience, and emotion constitutes the natural emergence of mind.[132] As I argued in the preceding paragraphs, this development is highly likely after the animal life form appears: motility is impossible without sensation and desire, and all three require centralized self-control. It also defines the design space within which further evolution is likely to take place: once motility evolves, together with sentience and emotion, none can be lost without losing the rest, and the likely evolutionary path becomes constrained by their interdependence. Phylogenetic retreat toward slower metabolism and losing sensation, emotion, and motility is possible yet improbable: the niche is already occupied by plants and fungi, and to survive, the newcomers will have to display some sort of advantage over them, which is not likely to be gained, especially given that the animal organism lacks the means to extract energy from its immediate surrounding by absorbing nutrients from the soil and using solar energy to process them. This sort of ability is not easy to evolve, and a "hopeful monster," an individual that, due to a lucky mutation, possesses some advantageous characteristic that lacked in any shape or form in its parents, is rather unlikely to be able to successfully reproduce.[133] Staying more or less the same over generations in terms of motility, sentience, and emotion is fine as long as no changes that endanger survival are present. These do not have to be major alterations of climate or appearance of capable predators, multiplication of similarly abled animals who compete over the same resources would endanger survival as well and require individuals to demonstrate superior abilities or perish before having a chance to reproduce. Evolving toward greater sophistication, on the other hand, is not only always possible but also seems advantageous in terms of survival and reproduction: it brings more capable animals to a new plateau where they lack serious competition and thus are likely to have better chances to survive and reproduce. Since the main characteristics of the animal life form are self-control with motility, sentience, and emotion, it is likely that the evolution would involve these characteristics. In the long run, this development seems inevitable barring a planetary catastrophe that would wipe out all life and the chemical precursors for its future re-appearance.

There are multiple ways to evolve in terms of self-controlled motility, sentience, and emotion. Of these, two particular aspects are important to the development of choice and ultimately human freedom: the aspect of emotional development and the aspect of social development, specifically that of signaling.

Urges, Instincts, and Emotions

The terms 'urge,' 'feeling,' 'emotion,' and 'instinct' are frequently used as synonyms or at least are presumed to have overlapping sematic fields.[134] However, for our purposes it would be more prudent to use them to refer to different phenomena that are important for understanding the evolution of freedom.

Any animal organism has urges – without them, as it has been argued earlier, it would not be capable of moving toward a goal set by the organism itself, and thus would not be able to acquire nutrients and survive. The essence of the urge is in the directionality toward its fulfillment: its early evolutionary role is to bridge the gap between the organism and the food it needs, to push the motile animal to move toward food. As such, the urge makes sense and is, in fact, necessary only in those life forms that are equipped with sentience and motility, i.e., animals. This basic essence the urge shares with emotions and is usually analyzed as such.[135] However, differently from Aristotle's appetitive faculty of which desire, passion, and wish are species and which is related to pain and pleasure,[136] the urge does not have to be related to pleasure and pain. It can be present in organisms simply as a push to do something, most commonly – to move toward a source of nutrition. We can see plenty of examples of urges that are not related to pleasure and pain among insects. Insects are not known to feel pain, at least in the mammalian sense of feeling pain: they do not tend to pay any significant attention to an injured body part, nor they seem to possess enough of the neural circuity that is related to pain responses in other animals who are usually thought to feel pain.[137] And yet it is hard to deny that insects have urges. Heidegger cites an experiment where a bee with a cut abdomen continues to gorge on honey indefinitely, since its abdomen is not getting filled up and therefore does not "switch off" the feeding urge.[138] This provides a good demonstration of what simple urges are: mere impulses toward something that are capable of initiating and sustaining behavior that is consistent with the impulse, at least in the absence of factors that would invoke other, more powerful impulses.

Urges do not require advanced cognitive abilities, if they require any at all. There is no need for memory of images encountered through various senses, nor for abilities like generalization over such images. All what is needed is sensation and a genetically determined tendency to behave in a certain way when a stimulus that triggers the tendency is registered. This is why urges are characteristic of all animals, even those that are neurologically simple.

Once the genetic infrastructure for urges develops, it evolves to be utilized for various purposes, i.e., to create various urges. Besides feeding, lust is present through much of the animal kingdom, including its biologically and behaviorally simpler quarters: an urge is necessary in order for an animal to get attracted to a potential mate and to perform the necessary sexual motions. In the same way one can also think about avoidance response and more.

Urges, however, are of little help in more complex situations: they merely push the organism in some genetically determined direction, given the right sensory trigger. Yet when what is needed to achieve the goal is more than mere movement but rather a sequence of behavioral steps, more than a mere urge is required. An urge is enough to move toward a smell, yet not enough to gather twigs and build a nest way up a tree so it is more conducive to producing an offspring that would survive for a while; it is also insufficient for lying in wait and chasing the prey once it is within range.

With the evolution of the neural structures and the elongation of the DNA that is required to produce them, the door opens for the development of instincts. Much has been said about instincts. The idea of having genetically determined responses to environmental stimuli not only seems appealing to us as an easy way to explain complex phenomena in the non-verbal as well as verbal animals, it also seems to match innumerable observations where animals with little demonstrated capacity to learn and no observable learning experiences to speak of demonstrate complex behaviors. Bees and ants have something that seem like an insect analogue of sophisticated social hierarchies, many birds sing complex melodies and perform elaborate mating rituals, etc. – the sort of activities that humans should learn long and hard to perform with a similar level of efficiency. The problem with instincts is that they are frequently defined in a way that confuses between them and urges, thus obscuring the differences that are important for understanding the evolution of freedom.

For example, Mary Midgley treats instincts as "genetically fixed tendencies" and, more vaguely, as "innate determining elements."[139] Different behaviors are summed up under this broad category: human crying and suckling, baby kangaroo's journey to the pouch, sexual behaviors, bee dances, bird songs, and more.[140] However, as Midgley notes, this "genetic programming" sometimes defines a whole behavioral sequence, what Midgley calls closed instincts, and sometimes leaves gaps to be filled by experience, e.g., by the imprinting where the first organism to come before the ducklings will be perceived by them as somebody to follow, or the songs heard by young birds will partially define their tune – the open instincts. Midgley notes that "the more complex, the more intelligent creatures become, the more they are programmed in this general way (with open instinct – MY), rather than in full detail."[141] Midgley's motivation is to counter those who, like Heidegger, argue that human beings are not animals at all,[142] part of the argument being that humans, having rational mind, do not have instincts. Humans certainly have what can be deemed, together with Midgley, natural tendencies, linked to our particular embodiment and rooted in our genetic endowment. However, the difference between Midgely's open and closed instincts is essential and does not warrant the use of one term.

General tendencies are the simpler form of urges. They do not define sequential behavioral patterns that are required for their satisfaction, neither in biologically simpler nor in more complex animals. The "general ruling motive"[143] like getting home or seeking water does not prescribe what to do – this is up to the other factors, be it the only thing a certain creature can perform, e.g., crawling for newly hatched turtles, or more complex, like learning how to hunt for large cats. Instincts, on the other hand, prescribe a sequence of behavioral steps.

Dennett cites the research on the digger wasp, *Sphex ichneumoneus*, that carries a complex multi-stage behavioral pattern focused on ensuring that its posterity has enough food to get by until they can explore the world on their own.[144] The mother wasp paralyzes her insect victim, usually a cricket, drags it to the threshold of a hole where she deposited her eggs, climbs into the hole to check on them, and then drags the future food in. If one moves the cricket a few inches, she does not drag it to the hole, but first brings it to the threshold again, climbs into the hole, and only then brings it inside. The last point is crucial: instinctual behavior consists of a set of defined steps that is triggered by a certain stimulus and performed as a whole. Its textbook definition is that of a fixed action pattern that is fully functional the first time it is performed, requires no learning and no previous behavioral experience of the cues that set the pattern in action.[145] Instinct is not a general tendency but a specific set of actions.

It is easy to see how instincts are useful for survival. An animal born with a mere general tendency to lay eggs might well lay them in a place that would not be appropriate for hatching. On the other hand, Dennett's devilish digger wasp, equipped with such an elaborately murderous instinct, has higher chances to have its genes passed on. Yet, given such sophistication, instincts do not require advanced neural machinery: it seems biologically and computationally easier to encode a sequence of steps than a large number of possible conditions and respective responses, not to mention the ability to learn and then remember what has been learned and have it ready for retrieval. In terms of energy consumption, the fact that instincts do not require learning is a big plus: learning requires time and resources, while worker ants are born without the need to be coached. No wonder that instincts are very common throughout the animal kingdom.[146]

While instincts seem to render behavior deterministic, this is where freedom begins, as I argued earlier.[147] The digger wasp has to make choices: it has to choose a hole to deposit her eggs, for example, as no perfect genetically transmitted prescription for a specific hole could be produced, only some perceptual specifications. She also had to choose which cricket to catch. Surely, these are very limited choices, and the digger wasp does not seem to possess any serious neurological structures for analyzing alternatives. Yet the main components of choice, i.e., access to represented alternatives, however primitively, seem to be present in it, albeit in a rudimentary form.[148]

Despite their advantages, instincts can also be detrimental to survival, particularly in situations of change. To continue with the example of the digger wasp, its instincts work as long as there are crickets around and no competitors that will routinely snatch them from the wasp hole's threshold while it is checking on her eggs. In fact, if a clever biologist removes the antennae of the insect victim she had just prepared for her future hatchlings, she is at a loss and would not drag it by its legs; the wasp would attempt to seize it by its head, fail, and leave the prey. The degree of freedom instinct allows, both by dictating behavioral sequences and availing cognitive resources to go beyond it, is extremely limited in the digger wasp's case. Therefore, a minor change in the availability of the exact prey it is after might well cause the wasp's extinction. This equally applies to other animals who are instinctually wired for certain food or shelter. This is why Dennett suggests that variable environments favor incomplete, or more plastic, design.[149]

This seems to be the reason for the "loosening" of instincts the higher we climb up the evolutionary tree in terms of physiological sophistication. Rats would have no issue consuming all kinds of food. Their instinctual endowment is much less specific and thus much less restricting than that of insects. It is their cognitive capacity that constitutes the main means of survival. This equally applies to many other animals. With nest-building birds, moving a twig a few inches would not matter. Goats will look for patches of grass everywhere they chance. Beavers can use various kinds of wood, and their dams are not the same. This increasing sophistication is enabled by the development of the central nervous system, physiological structures necessary for controlling an organism's behavior – choice is, in biological terms, self-control. It will also lead to lowering the contribution of instincts to survival.

This way, somewhat paradoxically, with the progress of evolution instincts fade and urges bounce back. In higher mammals, e.g., dogs and primates, we see some vestiges of instinctual behavior, e.g., fetching or sucking, yet nothing resembling the fixed action patterns of the digger wasp, ants, or bees. However, there is an important component in higher animals' urges that seems to be absent or is at least not that pronounced in insects: pain and pleasure. This makes the role of urges quite different and thus warrants a different term, emotion, an organismic function that in terms of evolution starts with the feeling of satisfaction and dissatisfaction accompanying varieties of physical pleasure and pain and then can be triggered by any feeling of satisfaction and dissatisfaction, whether they involve pain receptors or conceptual thinking.

Like urges, emotions play the survival role of enabling the animal to bridge the gap between need and satisfaction, to traverse distances over time. Yet basic urges, as present in less sophisticated animals, can do without pleasure or pain, satisfaction of the organism rather than "switching off" the urge versus dissatisfaction rather than keeping the urge "on." Urges originate within the organism, yet they do not necessarily involve the relation of the organism to itself. A bee does not necessarily feel pleasure when its abdomen is full – its full abdomen triggers the organs that are involved in the intake of honey to stop their activity. The bee also does not seem to recognize its own modifications – otherwise it probably would show some sort of behavioral change after its abdomen is cut; we have no reason to believe that bees feel pain or pleasure. We have little basis to claim that the bee relates to its own self in a meaningfully aware way. On the other hand, an animal that feels pain and pleasure does relate to its own self, has itself for an object just like it has external entities for objects of sensation.[150] The motivation in emotions is to avoid felt discomfort and to feel comfortable. The urge to feed, or to mate, or to care for another, aka love, or revulsion[151] are all "feelings accompanied by pleasure or pain."[152] It is this dichotomous feeling of pleasure and pain[153] that provides the motivation to go ahead with following the urge in the case of animals endowed with emotion. The sexual urge in mammals, for example, creates a discomfort that is alleviated only when the urge is satisfied; and so does the urge to feed. Love, an emotion that pertains to the "urge to care for another and make its welfare one's own,"[154] derives joy from pleasing this significant other and from her pleasure, is about feeling deep discomfort when such urge is not satisfied and comfort when it is. This makes emotions particularly suitable for animals who can exercise choice: the pleasure vs. pain calculus is, perhaps, the first criterion an animal can use when comparing alternatives.

Animals can have emotion even if they are not endowed with discursive metacognition and thus cannot consider the ways of making themselves feel better as ways of doing things that can be judged by criteria ranging over all of them. No non-verbal animal can compare the merits of the life of crime vs. studying and teaching philosophy as different methods that have a variety of implications to the organism's well-being. Yet none of this is required for the motivation to alleviate discomfort to trigger action.

In comparison to urges, emotions are a very efficient means to make sure that an animal bridges the gap between a felt need and its satisfaction. Urges, like instincts, are particular; emotions, on the other hand, are general, each emotion can be triggered by a variety of situations. To bridge the gap between a need and its fulfillment with an urge, it should be specific enough to relate to that need, e.g., activating and deactivating the organs related to feeding in a bee. It is evolutionarily taxing to provide an urge for every need and impossible to have urges that would match the environmental changes that take place during the animal's lifetime. With emotions, it is not important where the pain of dissatisfaction comes from – it will still trigger an emotion that will motivate the organism to alleviate it. This is the crucial distinction between urges and emotions. Bee's urge to feed comes when a specific bodily structure registers particular neural impulses, e.g., the lack of pressure in the abdomen. As long as this lack of pressure is present, the structures responsible for feeding operate. Once it disappears, they will stop. The emotion of dissatisfaction, on the other hand, can come from hunger, pain, lack of attention from other animals – potentially unlimited number of sources. This emotion is not connected to a specific bodily structure that is destined to satisfy it – instead, it will push the animal organism as a unity to seek satisfaction. The emotion is decoupled from the coping mechanism, and this makes the animal actively look to satisfy it. The animal will rely on its mental contents to find a way to satisfy the emotion: on associations acquired in the past, on cognitive maps, etc., depending on its level of cognitive development. Trial and error might follow, with learning resulting from it; future experiences might well change the relevant mental contents.

Thus, emotions constitute a much more flexible means of survival than urges. The emotion of anger provides a good example here. A strong feeling of displeasure and hostility, it can be connected to a variety of more or less generalized sensory images and thus invoked when similar conditions arise, motivating aggression or at least a display thereof – which can be useful in repelling enemies and competitors that endanger the animal's survival or procreation. As such, it can be triggered by events that the animal experiences during its lifetime and that could not have had genetically "pre-programmed" responses. Later, with the development of discursive intelligence in humans, anger can be invoked not only by retained images but also by thoughts.[155] Fear, aversion, sympathy and other emotions are similar in the way they function.

The pain-pleasure calculus is not straightforward, of course, since different emotions can compete and cause different degrees of satisfaction and dissatisfaction. As has been explained earlier,[156] arguing that there is a simple algorithm that decides which urge wins based on, say, the arithmetic relations between pain and pleasure is highly problematic. There is no reason to assume that non-verbal animals can non-conceptually measure these on a specific scale where the joy a dog derives from the approval of its significant human can be directly compared to the joy it experiences from eating a treat; and indeed, different dogs opt for human approval or for a treat in similar situations.[157] Since dogs possess representing consciousness, their opting for this or that alternative can be best explained by choice.

Plasticity

As was noted earlier, all life is plastic. As living organisms derive their sustenance from the environment, at least their composition changes. The prevalence of this or that chemical makes the biological processes in unicellular organisms intensify or slow down, and these changes impact multiple processes that help the organism maintain itself. In some organisms, e.g., slime mold cells, it can lead to rather complex changes. In plants and animals, plasticity takes different forms as they experience different evolutionary pressures.

Plants, as explained earlier, are morphologically plastic – they grow more or fewer roots and branches depending on the availability of nutrients and sunlight, for example. In animals, a minimal degree of morphological plasticity remains: differences in the availability and type of nutrients can lead to the growing of more or less hair, longer or shorter limbs, and larger or smaller size of the animal itself. Muscular plasticity can also be seen as morphological: exercising this or that muscle frequently leads to its enlargement in animals. However, the number of organs, including extremities, does not change: the animal is a more unified organism, as has been noted several times. Animal plasticity is mostly behavioral.

With the increased sophistication of a motile, centrally controlled life form, better adaptation lies in the ability to respond to different situations in a way that is conducive to survival. Specific adaptations might be the first step: the organisms that survive till reproduction are those to have the ability to cope with the situations they encounter. This *ability to* do something, e.g., to be of the right predisposition toward fighting vs. work to become a useful member of an ant colony, to digest swallowed insects, or dig holes and hibernate during the winter months, is a specific adaptation to a specific range of environments. Instincts can be seen as advanced *abilities to*. Yet, as has been argued earlier, they are also limiting, specifically in the situations where the environment is variegated and can change significantly enough to endanger the organism. Thus, *abilities for* emerge.

Certain physiological structures that enable the *abilities to* do something specific must be present in order for them to evolve toward greater sophistication. Specifically, the carrier of instincts needs memory and the ability to compare sensory output to stored images – otherwise it will not be able to select the right blade of grass for its nest, the right insect to put its eggs in, or the right figure to follow.[158] For instance, the ant has to have some sort of memory to return to the ant hill at night; the weaver bird needs to remember where its half-built nest is; and so on for all instinct-laden animals. These abilities are supported by the nervous system, and the more sensation and memory develop, the more sophisticated these structures become: those to survive and reproduce are the ones with the more effective senses, memory, and the coordination between them. *Abilities for* capitalize on these structures and utilize them for a new purpose.

Dobzhansky, in addressing the biological basis for human freedom, notes that "an essential feature of human evolution which has made our species unique has been the establishment of a genetically controlled plasticity of personality traits. This plasticity has made man educable and has made human culture and society possible."[159] Before enabling human culture and society, however, this sort of plasticity appears in non-verbal animals in the shape of the capacity to engage in a variety of different activities, *abilities for* developing ways to adapt to hitherto not encountered situations, situations for which the organism does not have an instinct to deal with. The ability to learn is the primary example of the *abilities for*, and the most important one for the evolution of freedom.

The very same neural structures that enable the processing and coordination of sensation and movement through storing images, once sufficiently developed, can be used for learning. At this point in the evolution of animals, they store the connections between sensations, thus enabling the retrieval of feeling on encountering the image with which it had been associated. This is achieved by neuroplasticity, a change in the nervous system in response to experience that affects its function and structure.[160]

As Costandi notes in his recent review of the research on neuroplasticity, "nervous systems evolved to change."[161] The research on neuroplasticity has hitherto involved mice, rats, cats, songbirds, macaque monkeys,[162] and humans, and it demonstrates similar processes by which neural infrastructure enables learning in all these species. For various animals, the neural system is based on neurons, special cells that can transport and process electro-chemical impulses of different kinds. The basic structure of the nervous system is formed during the embryonic stage based on what is encoded in the relevant segments of the DNA. However, the formation of new synapses, or connections between neurons where the minute gap separating them is conducive to transferring the neural impulse, also takes place following the exposure to certain stimuli. This synaptogenesis continues throughout the lifetime of the animal organism, even though the majority of synapses are formed during the embryonic period and the early age. The synapses can also be strengthened or weakened, and other cells, most notably the glial cells that surround the neurons, can be produced or extinguished – all in response to touching, hearing, tasting, smelling, and seeing things, as well as feeling certain feelings.[163]

We can distinguish two types of neuroplasticity: functional and structural. With the functional plasticity, neural cells can start releasing more or less of the neurotransmitters, chemical compounds that excite the next neuron or the muscle that is activated by the neuron, following certain experiences.[164] Also, following experiences, the identity of certain neurons can change: they can start releasing new neurotransmitters and thus switch, for example, from exciting neural impulses to suppressing them.[165] Structural changes, on the other hand, take place when the volume and the connectivity pattern of the nerve cell connections change. Both can occur on vastly different timescales, from millisecond to years. They can also occur following a brain injury. In these cases, remapping might take place, when brain areas that usually process the inputs from an injured sense organ or control the injured limb shrink, while those related to the still functioning sense and motor organs take their place. This phenomenon has been observed in monkeys as well as humans.[166] Addiction also involves changes in neural pathways.[167]

It is important to note that both functional and structural changes in the nervous system result not only from passive exposure to stimuli but also from active engagement with the world. The feeling of self, e.g., bodily effort, can be stored in memory just like a sight or a smell – somatosensation is as much a sensation as any other sensory output.[168] This way, tool use, characteristic of some non-verbal animals like corvids and apes, is found to "rewire the brain" by changing synaptic connections.[169]

An interesting corollary to the workings of neuroplasticity is the increase in individual differences. As noted in the discussion on individuation, animals endowed with learning become increasingly individuated by their life history. It is neuroplasticity that makes this possible. The individual history of exposure and response to stimuli now makes different animal individuals behave differently in response to their environment. Moreover: individual brains differ in their degree of plasticity.[170] Some of this can be attributed to the unique genetic makeup of the organism, yet some – to its history in its Umwelt. It is highly plausible that certain configuration of neural networks would be more plastic than others. Given that neural networks form as a result of behavior in the world and exposure to various stimuli, we can see how behavior impacts the range of abilities.[171]

Individual differences between animals can express themselves as differentiated mental strength. Some animal individuals would be capable of storing and associating more images, associating them along more lines of similarity, or performing their comparison with the intuited images quicker. As a result, some animals would perform better at navigating mazes, recognizing individuals from their and other species, etc. This may result from genetic endowment, exercising certain skills throughout their life histories, or both. All this is likely to make a difference in the lives of animal individuals.

Relying on neuroplasticity, the ability to learn gives rise to the transmission of habits, intra- as well as intergenerational. The most famous case here is, perhaps, the spreading of the habit of sweet potato washing by Japanese macaques on Koshima island.[172] Once started by a young female, the habit was imitated by other monkeys of her age group who observed her exercising it – presumably, they found it enjoyable. Then their mothers followed suit. The last ones were the males. Then it spread to the next generations and became what is deemed by de Wall a social tradition: there is nothing instinctual in washing sweet potatoes, there was a particular point of onset, and it is not practiced by other colonies of Japanese macaques. There are multiple other examples of habits started by a certain ape being imitated by others, from trotting to the same rhythm around a post to sticking a straw of grass in an ear.[173] Given an uncanny resemblance to the way human traditions are started and transmitted, de Wall and other consider this to the development and transmission of culture, defining culture as the "learning of habits and traditions from others, with the result that groups of the same species behave differently."[174]

However, there is a principal difference between the social transmission of behaviors and human culture. Human culture transmits not only behaviors but also the meanings behind them, meanings that relate mostly to the value or significance of behaviors. Humans sacrifice to the gods to appease them, march around the poles to emphasize their readiness to die for dear leaders, and put various objects in their hair in order to look fashionable. Every action of this kind is a subject for explanation and questioning, or normative evaluation – which requires thinking about behaviors in light of certain criteria. This seems to be inaccessible to animals who lack linguistic intelligence. Normative evaluation requires entertaining what is universal and relating it to what is particular, and this is impossible without discursive metacognition. Thinking about specific behaviors in light of criteria that can apply to any behavior cannot be carried out without understanding the relation between the universal and the particular, category and its instances, and such understanding cannot be achieved without the mind's reflexivity, metacognitive aboutness. What is accessible to animals capable of imitating each other is mere aping, pun intended. For people, this is a starting point, and every member of the human species endowed with a basic level of discursive intelligence can ask the question "why we are doing it?" That, of course, if they choose to do so.[175]

Neural plasticity has a very significant impact on the evolution toward freedom. Once its germ appears, it changes the equation of survival: a sophisticated *ability for* would be more useful in the long run than even an extremely elaborate instinct because the *ability for* is general, it enables the organism to cope with a wider spectrum of situations. Thus, the evolutionary pressure for species endowed with neuroplasticity would be to increase it, to develop further the biological infrastructure that enables plastic behavior, and first and foremost – learning. The more generic these structures are, the more pliant they are, the higher is their capacity to process stimuli – the better. And with a more pliant neural system, choice, the ultimate behavioral plasticity, becomes possible.

Choice

Retaining sensory images, which is achieved by the representing consciousness that is equipped with memory, is a prerequisite of choice in an aware self: to choose between alternatives, the self should have access to some sort of specification of these alternatives.[176] Retaining sensory images, or memory, is evolutionarily achieved by the quantitative increase of the basic neurological structures – in the case of life on Earth, the interconnected neurons. Humble rotifers, some of which lead a motile life while others – a sessile existence, have close to 200 neurons.[177] A jellyfish, lacking brain yet sporting a nervous system, has some 5,600.[178] Ants have around 250,000,[179] while naked mole rats, mammals that lead an ant-like existence in terms of the organization of their society – 26,880,000.[180] Chimps, our closest relatives, have twenty eight billion neurons, of which 6.2 billion are concentrated in the brain; this is fewer then orangutans and gorillas have.[181] Finally, humans have on average eighty six billion neurons in the brain, with sixteen billion in cerebral cortex.[182] This is less than African elephant's brain with its 257 billion neurons.[183] Based on the paleontological record and the relatedness of the currently existing species with the species it uncovers, we can assume that the number of neurons in the body is roughly correlated with how late the animal evolved. As we can see from analyzing the behavior of various animals, it is also positively, even though not perfectly, correlated with the animal's ability to retain and recall images and act upon their associations. However, quantitative increase is not everything: the way neural structures are organized seems to matter as well. More efficient neural networks that utilize smaller number of neurons to achieve the same outcome, e.g., storing images, retrieving them, or comparing intuited images to the recollected ones. This might free remaining neurons to handle more tasks, e.g., relate to own mental contents and the connections between them, develop the capacity for discursive metacognition, considering its own consideration of mental contents. At less ambitious levels, it can mean associating images by multiple characteristics, enabling the development of keener senses that supply more information, or making a signaling system more sophisticated.

Associating images, as it has been argued earlier, provides the future dimension to recollection: by linking a recollected image with the intuited one, an animal essentially makes a prediction of what is to come as a result of the state of affairs reflected in the intuited image: when seeing a patch of grass and connecting it to a kinetic response, i.e., walking forward, the animal retrieves the gustatory image of enjoying the said grass, and thus chooses to walk toward it, propelled by the feeling of hunger. This is how intentionality becomes teleology at the individual level.

Choice makes perfect evolutionary sense: it enables more effective responses to situations for which the animal does not have a well-defined instinctual response triggered by one of the situation's perceived aspects. Therefore, an animal equipped with the capacity to choose would have a chance to survive in circumstances where an organism lacking in choice would be likely to perish: the latter would have no response to them, while the former organism will try and associate intuited and recollected images, then compare the latter to what is aligned with its desires, including that of avoiding pain, and act accordingly. As noted earlier, this does not mean that animal behavior is determined by the comparative strength of desires or of urges – the possibility of such direct comparison on one scale common to all desires or urges is a philosophical fiction.

Since choice is evolutionarily advantageous, once it appears in its most basic form, it becomes a factor in evolution: animals who can choose better have a higher chance of survival. The quality of choice here refers to the breadth of associating images and its speed. An animal that can associate smell and various elements of visual image, e.g., lack of movement, is more likely to distinguish between a corpse of a predator and a live predator than the one for which cross-modal recollection is problematic. Similarly, an animal whose speed of response is faster has a higher chance both to catch its prey and to escape its predators. Therefore, neurophysiological structures and the abilities they support are more likely to evolve if they support choice rather than if they have nothing to do with it. An evolutionary escalation that was mentioned earlier takes place here as well: choice is enabled by certain physiological structures, and once it appears, it makes these structures ever more evolutionarily useful. They, in turn, enhance the reaches of choice – which leads to the survival of those who possesses the best physiological choice enablers.

There is another reason for the evolution of choice to progress once it shows up – animal interaction. Organism's sophistication needs to be commensurate with that of its food. Thus, organisms that fed on bacteria, plants, and fungi in the period that preceded mass predation did not demonstrate much sensory and neural sophistication: in the pre-Cambrian period what has been deemed the Garden of Ediacara was inhabited by slow-moving creatures without complex sense organs, "relatively self-contained and self-possessed beings."[184] Yet once evolution started to implement the patent of feeding on other animals in the early Cambrian period, they became very relevant to each other's survival, both as prey and as predators. From thence on, the Umwelts of living organisms started to intersect intensely in a new way, and the development of ways to survive while escaping predation or thanks to its success became urgent. Senses, motility, and the coordination thereof are the principle ways of animals to stay alive both as potential prey and as predators – and these became subject to evolutionary development.

Once choice widens the repertoire of responses and thus makes them hard to anticipate, the most economical way for another species equipped with the neural system of some sort is to capitalize on it and develop choice as well: to run after the prey or to drop the chase, to run away or to face the enemy by summoning the rest of the herd, etc. This, in turn, will give an evolutionary reason for the animals interacting with it to enhance their capacity for choosing – by evolving better neural enablers of choice.

If choice is evolutionarily advantageous, we should see its development and the corresponding development of its enablers in different branches of the tree of life. Indeed, this is the case: choice is a trait that evolves in animals whose genotypes diverges many epochs ago. Mammals' and birds' common ancestor lived some 300 million years before our days, in the Paleozoic era.[185] Accordingly, they have very different physiology and lifestyle. However, both mammals and birds are capable of choice, and many possess advanced choice enablers in terms of memory, ability to form associations between remembered images, etc.[186] Even more strikingly, the vertebrate branch of the tree of life, the one that gave rise to fish, reptiles, birds, and mammals, split out of the invertebrate branch some 600 million years ago, during the Ediacaran bliss.[187] This way, the common ancestor of today's dogs and octopuses had no developed senses, most probably no retention of images, and was incapable of exercising any choice to speak of. However, both dogs and octopuses, together with few other cephalopods, can exercise choice and have mental structures that enable complex choices.

An octopus's body is very different from the dog's: it has eight arms and no bones. The octopus lives under the sea, interacts with fewer creatures than canines do, has radically different means of communication and a very different frequency of communication, etc. However, octopuses possess a large nervous system that relies on some 500 million neurons, close to the range of a dog's, even though it is structured differently: in the octopus, most of the neurons are in the arms rather than in the brain, composing a distributed system that still controls the octopus's body very efficiently. Octopuses have sharp vision supported by eyes somewhat similar to human eyes, even though they evolved in parallel – our common ancestor barely had light-sensitive spots. In terms of cognitive abilities, octopuses can go through mazes, navigate efficiently, and learn how to unscrew jars. They demonstrate perceptual constancy, i.e., the ability to identify an object as the same under different lighting conditions. Octopuses adapt quickly to new environments and show remarkable learning: for example, some learned that springing jets of water on light bulbs shuts the offending light off, and used this knowledge to their advantage while causing serious damage to the lab. Octopuses were shown to recognize individuals of their own species and individual humans; they can learn by watching others; and have a theory of mind – they behave differently when they are under the impression that they are being watched by their human captors as opposed to not being watched. Finally, octopuses demonstrate advanced tool use, e.g., by using coconut shells as a mobile shelter.[188] All this would not shame an advanced member of the monkey species. In almost all of these examples, octopuses and cuttlefish demonstrate the ability to choose between alternatives: you need one to pick coconut shells rather than oyster shells to be used as a shelter, to respond differently to different humans in different circumstances, etc.

Thus, choice is a trait that evolves in animals whose genotypes diverged many evolutionary epochs ago. It is, therefore, a convergent trait.[189] It is also an analogous one rather than homologous: different bodily structures that evolved in various branches of the tree of life underly the ability to choose between alternatives. This is because the logic of evolution pushes toward the development of choice and thus toward freedom: life harbors the germ of freedom, and many different environments make its development advantageous.

As any ability that is advantageous for survival, phylogenetically choice is likely to persist and progress, with the evolution leading to the increase in the scope of choice, the number of dimensions over which the alternatives can be compared, etc. Without the possibility to choose between alternatives, which requires the ability to represent them mentally as generalized sensory images, such an increase would be merely more data for a genetically fixed algorithm to work with. Yet, as argued earlier, nothing suggests that such a fixed algorithm is indeed at work. The pleasure and the pain associated with different alternatives are unlikely to be compared on some scale common to all situations, and no empirical observation or logical necessity leads us to the conclusion that such scale exists for an animal. It is also highly problematic to claim that an animal that chooses according to its desires is a slave to its feelings and thus cannot really choose. The desires are part of the animal, it is no more possible to separate the animal from its desires as it is possible to separate it from its body, or animal body – from its mind. A dog that chooses to follow the desire for approval from its human over the desire for a treat is making a choice based on its own mental contents; so is a dog, the same dog or a different one, that makes the opposite choice. Thus, "from one level to the next, [animal] freedom becomes less individually restricted and more general and universal. The levels of freedom are increasing."[190] However, the non-discursive life form seems to impose a limit on the development of freedom, a limit that is caused by what Heidegger and others referred to as animal poverty.

Animal Poverty

All animals have a world. As selves equipped with sensation, they register various aspects of their Umwelt and relate them to their own urges at the very least. The scope of the animal's world is defined by the reach of its senses and the abilities of its self-control system, e.g., the nervous system, to process sensory outputs. The world of Uexküll's tick is quite limited – its senses are focused on a rather small range of olfactory and thermal stimuli, it can do little with them, yet it is a world nevertheless. The world of an elephant is much wider: it has all six senses relatively well-developed, with hearing capable of registering infra-sound inaccessible to most animals, which extends its sensation by many miles and allows them to escape stormy weather ahead.[191] Elephants are capable of recognizing different human languages and altering their behavior in response to male calls in the language of those who might harm them, they have well-developed emotions, signaling system and social lives. Yet, according to Heidegger and his interpreters, animals are poor in world (*Weltarm*) and this is what differentiates their world from the human one.

Inanimate objects, argues Heidegger, lack the world (*sind weltlos*) – they cannot have a world, as they lack sensation. Animals, on the other hand, are deprived of world while they in principle can have one, and this is why they are *poor in world* – poverty means not having something that can be had.[192] No rock can experience the sun – rocks cannot experience at all. The lizard very much experiences the sun – it senses its warmth, relates it to its internal urge to stay warm, positions itself in such a way that it is exposed to the warm sun rays. There is a way in which the sun is given to the lizard – and this way is the one of sensory experience and its processing by the lizard's nervous system. However, argues Heidegger, the lizard does not know the sun *as* a sun. Non-human animals are incapable of grasping the factors in their Umwelt *as such*, do not relate to beings as beings, do not "*stand within a manifestness of beings*."[193] The relation of "as such" is inaccessible to non-verbal animals,[194] they cannot lift their minds above the images and the relations between them – they lack discursive metacognition,[195] the necessary instrument for this sort of cognitive feat. For Heidegger having a world "implies *amongst other things* the accessibility of beings *as such*," going "beyond mere acquaintance with something."[196]

Heidegger tries to further cement his thesis with the discussion on animal captivation (*Benommenheit*). Animal Umwelt, argues Heidegger based on Uexküll, is fixed. The animal is confined to its environment, presumably defined by its genetic endowment, to its *disinhibiting ring* – a set of environmental factors the animal can register. The animal is taken (*benommen*) by these factors, cannot escape the ring, is captivated by them.[197] It is incapable of altering the ring of its specific disinhibitors or breaking it, engaging with new environmental factors.[198] In his argument Heidegger relies on the research of insects, specifically the bees, and discusses at length their instinctual nature and their obliviousness to whatever their instincts ignore.[199] It is easy to find similar examples in other species.

However, it is also easy to find plenty of example that demonstrate that the ring of disinhibitors, the limits of animal's Umwelt are not nearly as rigid as Heidegger's bees suggest. The examples of animal plasticity brought up earlier belong to this category. Many animals learn new relationships between different elements of their perceived environment, e.g., through conditioning. Chimps can be taught to categorize photos according to the distinctions suggested by people,[200] develop affection and rejection toward other chimps and humans, and even learn hundreds of signs in the human sign language, thus communicating their requests that can get to the point of gesturing at a familiar human "Dirty Jack gimme drink."[201] Dogs and cats and bears can learn tricks, signing birds – new songs, etc. None of these behaviors are instinctual in the classical sense or the Heidegger bee's; all are about stretching the ring of disinhibitors.

Animals can relate things by connecting experiences to images of events and generalize over images, some – in fairly advanced ways. Mary Midgley strays quite far from Heidegger's conclusions and equates this with understanding, arguing that "*Understanding is relating*: putting things into a context."[202] So do many others, e.g., those who use the metaphor of artificial intelligence in its literal meaning. This, however, seems misguided. It is possible to build a machine that "classifies" visual images or sounds by finding commonalities. A computer program of the neural network type can process multiple shapes, extract common features, e.g., color, size, etc., and categorize them into classes that are not pre-defined. This, however, will give it no more understanding than a person who categorizes Hangul characters into those with a circle and without a circle has of the Korean language: this categorization, meaningful for somebody who knows that the circle refers to a stop, a missing letter in the usual three-letter combination, is as meaningless to our categorizer as the categories of "fox" and "rabbit" are for a machine that "categorized" images of foxes and rabbits based on shape and color. We have no indication whatsoever that the chimp who sorts images into those of people and those of animals has any understanding of the meaning of these categories: most probably, it goes by similarities; these can be the looks, the behavioral characteristics it picked up like communicating, or simply the meaning for the particular chimp – after all, these are the people who reward it for doing useless things. Understanding requires more than differentiating between X and Y based on their sensory features or features derived from their images, e.g., frequency – it necessitates considering the relation between X and Y, as well as between X and Y and their context. It is the knowing that the images of foxes and rabbits refer to animals, having a concept of animals and images and the relationship between them, that indicates understanding. Same applies to Hangul characters: without knowing that they express sounds, that they are grouped in threes, etc., categorizing them into those with a circle and those without a circle does not give rise to understanding. This requires the ability to go meta which is accorded by language.

Thus, Heidegger animal's poverty-in-world might actually be poverty in the world. Non-verbal animals have access to an impressive array of elements, can form sophisticated relations between them, and are capable of increasing the size of their Umwelt. Yet they are unable to go meta, to consider themselves as selves, the world as a world, the relations of themselves to the world as good or bad, rich or poor, in need of expansion or satisfactory as they are. Consequently, they are also unable to expand these relations in ways humans can, e.g., by altering the world with the hope to make their relations to it more agreeable. The animal, standing before the world, has it as open yet not disconcealed, not openable further.[203] It might link touching a fence and experiencing an electric shock, yet not ask the question "Why does the fence cause the shock?" It stands before the world "half closed in its openness, half open in its closedness, becomes a very ambiguous figure."[204] The animal cannot take a standpoint outside of its Umwelt, however expanded by learning. Non-verbal animals cannot have this standpoint because they cannot apprehend something *as* something.[205] Consequently, their freedom is as limited as their Umwelts, their choices are confined to whatever alternatives the world as it appears to them through the sensory images, remembered and related in more or less sophisticated ways, processed in light of the animal's urges and emotions, makes available.

Animal Communication

Repeating the ever-present evolutionary pattern of utilizing existing structures and processes for new purposes, animal communication starts with the intercellular communication characteristic of single cells. Unicellular organisms emit ligands, molecules that have a tendency to bind to other molecules. The case of the slime mold, cited earlier, demonstrates most clearly how this causes multiple cells to respond similarly to another cells' chemical output. In plants, we see only direct signaling across gap junctions where adjacent cells can communicate with each other by emitting and immediately absorbing ligands. In animals we start seeing the ligand-based communications traversing slightly larger distances between nearby cells: paracrine signaling that gives rise to synapses, the type of communications at the basis of the nervous system.[206] Additionally, animals develop endocrine intercellular signaling which allows distant cells to issue messages to recipient cells that can be farther down the bloodstream – the basis for hormonal communication, the foundation of many responses that involve the organism as one unit, e.g., adrenaline and its effects on then fight-or-flight behavior.[207] These two types of intracellular communication enable the organism to control itself as one unit, which gives serious advantage in terms of survival: now looking for food and escaping danger becomes a possibility. This, in turn, strengthens the chances that the genes underlying these more advanced communication abilities would propagate further and develop toward greater sophistication.

In many organisms we have a similar sort of communication as the only one available. Social insects emit pheromones on many occasions, e.g., when in the state of sexual arousal, when startled by something, when carrying food, etc.[208] The release of pheromones most probably does not involve any sort of volition or planning by an ant or a bee, it is a by-product of its physiological state. However, due to its survival value, the species members evolve to respond to this signal, e.g., engage in mating, prepare to defend the colony, or follow it to the source of food. This way a mere byproduct of animal functioning, a cue, turns into a signal, the ability to emit which becomes subject to natural selection[209] – the process that exists throughout the communicating animal kingdom.

With the evolution of animal behavior and cognition, the involuntary emission of signals and responding to them by the receiving organism continues in those areas of behavior that are either not subject to voluntary control or are not consciously attended by the agent. For example, pheromones are thought to be responsible for the coordination of menstrual cycles between cohabiting women.[210] Similarly, we tend to respond with a startle to sudden movements of others. However, new ways of communicating evolve together with bodily types, behavior, and cognition.

With the evolution of the ability to store and associate images, animals become capable of associating events and feelings, e.g., a certain facial expression and its likely consequence. This way, bare teeth, a widespread sign of aggression in mammals, can become the means to show aggressive intent. This is a crucial development in animal communication that seems to become a factor in natural selection. When a canine attacks, it naturally bears its teeth: this is because teeth are their attack weapon. Other animals learn to connect between bare teeth and being attacked, either from observation or from their own experience. Given the image of the results of the attack present in the canine's mind, it can evaluate it, doing its pleasure calculus where the relative size of the attacker, its remembered history, etc., figure as factors. Such evaluation does not require verbal cognition, merely comparing stored images and, perhaps, some instinctual baggage. As a result, a canine beholding the bared teeth of its opposite number makes a choice – to confront it or to retreat. The results are stored as an image too; if, of course, the attacked survives the encounter. They also become part of the mental store of the attacker who learns that baring its teeth elicits fear in others. Given the ability to store, generalize, and retrieve images, it becomes part of its behavioral repertoire: now, having the intent to elicit certain response in other animals, it will bear its teeth, and this behavior is now connected to certain responses from other animals.

Here, just like in the case of ants and their pheromones, we see utilization of a biological process for communication purposes. Yet ants are not known to retain sensory images, generalize over them, link them together based on similarity in this or that sensory dimension, and compare stored images with the sensed ones – their responses to pheromonal compounds are inherited with most, if not all other aspects of their behavior. Animals, on the other hand, have representing consciousness and thus choice and can intentionally generate signals to elicit certain response when they are motivated to do so by, for example, urges or emotions.[211] We can see this process not only in felines and canines but on other branches of the tree of life, within species and in interspecies communication, e.g., between predators and their prey.[212] This is because the underlying natural selection processes are similar across the board, no matter what the biophysical infrastructure of signaling is.

One can say that the ability to connect sensory displays to mental images is indicative of animals' ability to produce symbols.[213] For example, vervet monkeys use distinct vocal calls to indicate the presence of a leopard, eagle, and snake – calls that elicit distinct response, as there is not point to stand up and look around in the case of a leopard, while this is the right response in the case of a snake as far as survival is concerned.[214] Here, each call can be considered a symbol for the predator in question. While these calls do not bear physical resemblances to the animals they refer to, the calls themselves are of a natural origin, and their connection to the predators is not devoid of a physical link – after all, they are all alarm calls, a natural response to danger. Moreover – it is possible to teach many animals to use symbols, e.g., vocal expressions for parrots[215] and hand gesture to chimps,[216] to refer to things and feelings. Once familiar with the signs and the responses they might elicit from those who know them, the animals would use them to ask for things, express their disappointment, etc.

However, there is one important limitation that is observed in animals that can use symbols – they never produce them on purpose. Chimps can be taught to use hundreds of words in the American Sign Language, yet in its nature their communication, while quite extensive in terms of the number of gestures used, is more like that of other animals: bodily postures of dominance and submission that trace their visual expression to behaviors like attacks, natural change of plumage indicative of sexual arousal, etc. This is because non-verbal animals lack discursive metacognition, the ability to consider cognitive relations, to think about thinking. Consequently, they lack the understanding of the nature of symbolic relation – such understanding would require having as an object of cognition the link between a symbol and what it symbolizes, thinking about a symbolic relation. The symbolic relation itself lacks a sensory component, it is not an image but a relation between images, a relation where one image stands for another image or for a category of images that share something that can be sensed. Therefore, non-verbal animals do not produce symbols with the intent to symbolize; they can produce symbols accidentally, e.g., by connecting specific calls to the images associated with danger or food, and then respond to them and use them – this does not require understanding the nature of symbolic relationship. For the same reason animals do not produce signs: producing a sign requires understanding of the relationship of significations, where there is no shared sensory basis between a sign and what it signifies. Whatever signs they are taught they use as symbols, tied to their referents by remembered past experience of temporal precedence and spatial co-occurrence.

In non-verbal animals, communication has an important social function, intra- and well as interspecific. If not for the communal living of the vervet monkeys, they would never develop the system of signals to indicate the presence of predators: the survival value of crying out after spotting a leopard for an individual monkey lacking any means to scare it is sharply negative. It is the survival of the group as a collection of individuals, and thus the chance of other monkeys to pass on their genes, that benefits from it. In fact, in a-social animals signaling does not develop. For example, octopuses, highly cognitively capable creatures, change colors in response to what they experience, e.g., encountering an unfamiliar object, being startled, etc. These are most probably "inadvertent expression of the animal's inner processes,"[217] reflection of its neuronal activity. Octopuses have excellent vision and can easily pick up visual signals. However, since they keep the communication with their species members and other animals to a minimum, they do not develop signaling – fabulous colorful displays remain without a recipient.

Animals capable of choice are also capable of more advanced communication, i.e., creating displays intended to elicit certain response in signal recipients. This, because the same cognitive abilities, namely memory and generalization, enable choice and intentional communication. However, choice and communication are orthogonal in non-verbal animals. Their ability to communicate as such does not impact their ability to choose, and vice versa. The reason is that animal signaling is external, it targets other animals, its elements cannot be used yet for relating to the animal's own cognition. However, the capacity to connect images to events even when there is little physical resemblance between them forms the elements from which human language is later constructed. Similarly to the way the emission of ligands later evolves into pheromonal communication in social insects and natural postures give rise to visual symbols of aggression and submission in mammals, the ability to connect calls and gestures to feelings and events evolves into language. The latter, besides greatly enhancing information exchange, gives rise to a different kind of choice – human freedom. Chimps stand on the brink of it, and humans cross the threshold.

THE EMERGENCE AND EVOLUTION OF DISCURSIVELY INTELLIGENT ANIMALS

Humans Vs. Other Animals: Language and Metacognition

In terms of anatomy, the differences between humans and other great apes, especially chimps, are quite small, much smaller on any count than the differences between chimpanzees and macaques, for example. Since the anatomy is determined almost entirely by genetics, there is little wonder here:

humans share 99 percent of the genetic code with the common chimpanzees (*Pan troglodytes*) and with the bonobo (*Pan paniscus*), while differing from these two species in distinct ways, i.e., in different sections of the DNA. With gorillas (*Gorilla gorilla*) we share 98% of the genome, while 15% of human genome is more like a gorilla's than a chimp's or a bonobo's.[218] As described in the last section, many cognitive capacities are also common to us and our primate cousins: various emotions and the bodily expressions that typically accompany them, the ability to make tools, the ability to remember experiences, associate them, generalize over remembered images, and more. Starting with the ethological tradition that compares humans to other animals, we look at specific abilities and find that we share a lot with our non-verbal kin. In fact, ethologists find pretty much for every specific ability, be it empathizing with other animals, restraining urges to conform to the ways of the society, deception, or communication, that the difference between humans and other animals is that of in degree rather than in kind.[219] And yet it is hard to miss the fact that neither chimps nor bonobos argue about the similarity between their cognition and that of humans, nor do they conduct debates where each side marshals arguments to support their position regarding the proper ways to treat other animals. They might well have fear and trembling yet not religion; they might be acutely aware of the danger of hanging on thin branches, yet they have no physical science; they have power struggles yet not political discourse; etc. The way of life humans typically have is radically different from the other primates' way of life – this point hardly needs substantiation.

There is one important and easily observable difference between the behavior of humans and that of all other animals – language. While all mammals and many other non-verbal animals communicate, this communication is always image-referential: their signals are always linked to feelings or images. Different animals have signals, common vocal, gestural, tactile, or other ways to indicate dominance vs. submission or spotting an eagle or a snake. Many possess the ability to learn new signals, e.g., sign language gestures, and stringing these signals together, e.g., in issuing requests or referring to something. New signals emerge, other animals of the same or a different species can learn them: remember, linking to internal or external events, retrieve the signal when an experience similar to the one linked to it comes about, or retrieve the corresponding experience image when the signal is registered. However, nowhere in the non-human animal kingdom do we find "evidence for symbolic communication, equally rich and multifunctional as ours."[220]

Human language is frequently viewed as a way to transmit information between conspecifics. Indeed, one can argue that much of the use of language is for this purpose, and that the initial language acquisition proceeds through communication with primary caregivers. This is true not only of human language but also for much of mammalian communication and even for significant aspects of bird songs. From here, one can argue, as Lee et al. do, that human language developed from the interaction between hominids using vocal signals to communicate with each other.[221] This communication is rich and frequent enough due to the genetic predisposition of hominids for interaction, the *interactional instinct.* This predisposition is supported by the neurological aspect of affiliative interaction between infants and their caregivers, an interaction that provides reward in the form of the release of endogenous opiates, and by the ability that hominid children share with other primates to detect patterns in their environment.[222] When the communication is extensive, a structure emerges out of the chaotic richness of the lexicon. Such emergence is quite common: "complex adaptive systems are seen to emerge spontaneously from the interaction of large number of agents and/or large number of items," be it flight formation of birds, patterns of traffic jams, or ant colonies.[223] This is how grammar is born – no need for language glands. Lee et al. also argue that no "neural substrate" has been found for Chomsky's universal grammar, nor "has it been possible to conceive of a credible evolutionary scenario for the genetic basis of UG."[224]

The problem with this argument is that it misses a crucial point: while both humans and chimps communicate and pass information to each other, only humans engage in discussions about the way of passing information. Much of human communication is about the relations between things and various aspects of these relations, about the hypotheses regarding the nature of various relations and the validity of these hypotheses, about the permissibility of actions, and other subjects that focus not on what we feel, see, or want, but on the feeling, seeing, and wanting. The main point of difference between animal signaling and human language is not in the richness of the symbolic communication and even not in the ability of human language to "transcend [. . .] the here and now":[225] after all, when a chimp signs "gimme water" or a bird sings a mating song, there is a transcendence of here and now – an intent that is pointing to the desired future state of affairs. The main difference between the communication of non-verbal animals and human language is that the latter is not limited in terms of the subject of aboutness – it can be about anything, not only about images that are traceable to sensation. Human language enables us to consider relations between images and the quality of these relations. We ask questions like "Is the relation of size important?," "Do these phenomena differ in kind or in degree?," "Is this a proper thing to do?," and "What is the meaning of 'proper'?" Human language, no matter what specific language we are considering, enables asking an unlimited number of such questions and questions about these questions, or engage in an unlimited number of levels of inquiry; yet none seems to be accessible to even the most advanced communicators among other animals. This unlimited aboutness cannot emerge from an unorganized large lexicon of signals, as it is not a structural feature of communication. This aboutness is also expressed in grammar, specifically in its universality: no matter what specific grammar of which specific human language we consider, it always can support an unlimited number of levels of inquiry, asking, for example, normative questions like "is this good?" or a widely causal question, "why?," about anything. This is very different form the spontaneous organization of a flock of birds or an anthill – neither is universal or unlimited.

What underlies this ability to think about thinking, to "step out" of any context and ask questions about it? This has been addressed in Chapter 2, section *Language and discursive metacognition*, as a capacity for meta, or metacognition. Here I will try to develop an evolutionary account of the development of metacognition, its impact on the human way of life, and its role in shaping human freedom.

The Evolution of Metacognition

The pre-verbal mind can consider some mental contents, namely intuitions, representations, and their generalizations. For example, it can associate them with each other and then act upon encountering one as if it means that the other is present, as happens with conditioning. More advanced non-verbal minds can recombine remembered content and imagine new situations, e.g., by generating alarm calls in anticipation of other troop members running away and then using this situation to the cunning individual's advantage. Metacognition, the ability to relate to one's own mental contents, however, does not have to be limited to the type of content that the mind can address – in its discursive form, it ranges over any possible content, becoming a capacity to think about any mental representation.[226] Once such ability develops, the mind can consider anything: the relation between images, relations between these relations, different aspects of such relations, etc. This opens an unlimited scope before the mind: normative evaluation, where certain relations between events, images, statements about events, and the self are subject to appraisal according to some criteria; these criteria itself; models of the possible states of affairs and the wisdom in developing such models; laws of nature that are abstracted from the observed events yet are somehow primary to them; entities that are constructed as having no sensory content as part of their nature, e.g., disembodied divinity, immaterial soul, square roots of negative numbers, musings on the nature and basis of knowledge, etc. In a self-determined organism where the body controls itself through its cognitive aspect, this also opens a horizon for choice and action that is as wide as the reach of the mind.

Attempts to find this sort of metacognition in other animals yield very humble results: only three monkey species have been shown to have some traces of this capacity. By and large, this is a human asset, a relatively recent evolutionary development.[227] The survivalistic value of metacognition seems clear: with the ability to consider any mental contents of one's own, e.g., way of behaving and plans of action, in light of criteria that have been formulated relatively recently based on the analysis of the world, an organism is capable of avoiding certain behaviors that are suggested by its emotions or past experience and pursuing other paths. This is likely to lead to survival in situations where other organisms that are incapable of such feats would perish – novel situations where the emotional responses honed by the past evolutionary history would be detrimental to survival or even lethal.

Some scientists suggest that this capacity frees the organism "from the constraints of the stimulus."[228] This seems to be a somewhat lopsided view of discursive metacognition. Stimuli, intuited and stored images, still impose constraints – it does not seem that without them language-enabled metacognition would have any materials to work with, at least in its early stages.[229] What discursive metacognition enables its bearers to do is considering these images in a variety of ways, in a multitude of contexts, some of which are not derived from the sensory qualities of images themselves. As such, discursive metacognition has a thoroughly transforming effect when the way of life is concerned, creating a radically new life form.

Pretty much every cognitive capacity we observe in nonhuman animals and have reasons to believe to be handed to proto-humans by our evolutionary ancestors is altered by discursive metacognition. Primates have been found to possess a theory of mind, i.e., to imagine what might be the content of another agent's mind in regards to, say, a hidden object. However, this is not discursive metacognition yet: behavioral learning might be sufficient to grasp that seeing an object move in space is related to looking for it later. Discursive metacognition takes place when a psychologist thinks about such a capacity and plans new experiments to test it, and when a philosopher examines her conception of the theory of mind. This equally applies to deception: chimps might exercise it, yet evaluating it ethically requires metacognition, thinking about deception in terms of good and bad. Many animals, and certainly all mammals and birds, make choices while taking into consideration the laws of physics: flying in formation that uses the air stream from the leading bird, throwing rocks and using sticks to get bananas, and many other similar situations attest to a rather precise attunement of animal behavior to physical constraints of their surroundings. In the case of using sticks, Köhler's and de Waal's chimps also demonstrate the ability to learn how physical objects work rather than relying on instincts. However, no non-verbal animals conceive of physical laws as principles of nature independent of their will – none shows any signs of conceiving of them at all. This is because they lack discursive metacognition requisite for thinking about the concepts of nature and law. As a result, apes do not write physics textbooks, which are essentially meta-physical since they are about physics. Nor do they write metaphysical treatises or explore the status of what is.

What is required for discursive metacognition to develop? It seems like not much, or, perhaps, more of the same. With a well-developed ability to generalize over images, intuited and stored in memory, apes already have the neurological structures to consider mental contents; perhaps, in this capacity they are stronger than other non-human animals. All what is left to develop discursive metacognition is for this structure to start considering mental content other than images, e.g., relations between images – this removes the limitation of the type of content that can be considered, thus enabling going meta on anything that is represented mentally. Nothing suggests that this would require novel neurophysiological elements. Biological research finds remarkable resemblance between human and primate brains; in fact, mammalian brain in general, from that of a mouse to the dolphin's and the elephant's, shows great similarity in structure, alongside differences in size. The elements of the nervous system, peripheral as central, are also similar: the neurons humans have and the neurons octopuses possess are quite alike in shape, biochemistry, and functions. Form here, it is most likely that the sheer increase of the number of neurons and differences in their configuration might have led to the appearance of the ability to go meta on the mental content that has no sensory component by adding processing capacity and reconfiguring neural connections. This reconfiguration can be achieved by neural networks themselves – similar reorganization processes are observed every time animals acquire a new memory, for example, or form an association.[230] Alternatively, a mutation that led to freeing up some processing capacity of the brain might have availed neuronal resources to engage in considering more than images. An example could be automatization of certain processes, e.g., face recognition, that once required more neural resources yet thanks to an advantageous mutation started to require less at a certain evolutionary point. It also might have been a decline in the functioning of some sense that freed up resources. Dogs have an extremely keen sense of smell and sharp hearing that much exceed those of humans, and they need significant brain resources to process olfactory and auditory sensations. We know that brains are highly plastic, and when some sense is lost, the brain areas usually dedicated to processing its outputs dwindle while the areas normally processing outputs from other senses encroach on them.[231] Perhaps, a genetic mutation handicapping the senses of smell and hearing led to freeing up the brain resources dedicated to them, and these re-organized into more complex neural networks, thus giving rise to metacognition. Or both adding more neurons and freeing existing resources were at work. These are, of course, speculations. Yet they show that there is nothing impossible in using the neural structures present in animals not capable of discursive metacognition to develop one. Moreover, it is likely that this scenario comes to pass in organisms with plastic central nervous systems capable of mutating: there is a high chance that the number of neurons available for supporting discursive metacognition will appear at a certain evolutionary point, and then one of the scenarios suggested above will take place.

Metacognition and Language

The capacity for metacognition not limited to sensory contents, just like any other ability of a living being, needs to be carried out to be of any use and thus survive the test of natural selection. The specific ways in which this capacity impacts behavior would depend on the evolutionary design space available for its exercise, or on the already developed abilities that can be re-purposed to support its instantiation. In other words, there had to be something that the hominids who evolved the germ of metacognition could do that would serve as its vehicle, just like chemical messaging between cells had to be there for the nervous system to evolve.

Such capacity in primates is signaling. The ability to connect between sensory images and memory and then generate certain related signals that impact the right sense in the right way in other individuals is present in apes, and there is little doubt it was present in the common ancestor of today's chimps and humans. Similarly to today's chimpanzees, early hominids most probably were capable of using symbols, acting upon the remembered connection between auditory or visual sensations and remembered images, more or less generalized. With discursive metacognition, though, the agent is able not only to use symbols but to address the link between the symbol and the symbolized, or to comprehend the relationship of symbolization. This is the crux of the evolutionary contribution of metacognition.

How would human mind address the relationship of symbolization? In the same way it addresses mental images and their generalizations – by assigning a label to them, something that refers to them and, once recalled, invokes them. The mental technique is already there, it is its application that is new: the capacity to go meta on everything, the newly evolved aboutness that can pertain to any content, sensory and non-sensory alike, uses the old ways of assigning labels. Yet now it is not a leopard, or its generalized image, that is labelled but the relation of assigning a specific symbol to the image of the leopard.

This unlimited aboutness of discursive metacognition has been construed by Heidegger, following Aristotle, as λόγος ἀποφαντικός, "discourse that points out,"[232] the way of thinking that reveals, or disconceals (ἀ-ληθεύειν) something that is not sensed. Such discourse seeks to point out the meaning, something *about* other things, something that is meant. To do that is to show something *as* something, or to take things together, to ap-prehend (*Ver-nehmen*) them. This synthesis is, as Aristotle argued, a necessary condition for the truth judgment[233] – and thus metacognition if the basis on which reason relies.[234] An important point here is that discursive metacognition as ability (*Vermögen*) leads to a possibility (*Möglishkeit*),[235] yet by no means it determines any particular instantiation, e.g., this or that specific way or shape of thinking; nor does it necessitate that the possibility will be realized at all. This is the umbilical cord that connects reason to human freedom: the scope of possibility for verbal animals equipped with discursive metacognition is not just wider than that of non-verbal animals, it is virtually unlimited. Which possibilities become alternatives for choice depends on the world humans build for themselves, as I will argue in the next section.

Once the relation of symbolization is understood, symbols can become signs, losing the aspect of sensory resemblance to what they symbolize.[236] This way, a hominid equivalent of the vervet monkey's call that refers to a leopard, while it might retain its original auditory qualities, starts playing a completely new role: it not only triggers the recollection of the sensory aspects of leopard's sighting, it is also understood now as *standing for* such image. With this understanding, the now linguistic agent can invent another symbol for leopard's sighting, a symbol that has nothing to do with the sensory qualities of the first one – a sign. It can also start using signs for a variety of mental contents, e.g., generalized representations, thus naming general representations, or creating categories – wild animals, people, food, trees, tools, etc., whatever she finds necessary to name in her environment. Other individuals who interact with her, endowed with discursive metacognition, can understand hers and suggest their own names. The understanding here comes from applying the recently evolved form of metacognition to the ability to link generalized images that has been there for an evolutionary while. The interlocutors of the individual that suggested the name understand the relationship of signification, the arbitrary nature of signs, and thus can easily suggest their own. Yet not only categories can be named now – anything can have a referring sign. This includes the relationships of symbolization and signification themselves. Since there is no limitation in principle for the number of signs to be used by linguistic agents, metacognition becomes generative: it enables people to invent as many signs and names as they see fit.

One aspect of language has been fascinating people probably since shortly after the dawn of history – the grammar, and particularly the differences in grammar between different languages. It is relatively easy to understand how different groups of people came to use different names for similar events – it is in the nature of naming to be arbitrary. However, the question of why Chinese and Ukrainian would have no articles while Arabic and French would seems different. Yet there is no evolutionary reason why it should. Specific grammar can be seen as a self-organizing feature of communication – it appears in pretty much every other area where different agents interact, from mating rituals and flight formations to traffic jams and the Internet. There is no reason why there should be only one way for symbolic communication to be organized. The only constraints here are comprehensibility for all the communicating parties involved and the ability to refer to the contents, sensory as mental, they want to refer to. The dynamic changes in grammar over the years, of which the English language is a prime example, attest to the ease with which different grammars can enable the utilization of the capacity to go meta on mental content. The ease with which various governments enforce "standard" grammar through schooling is another example of how the nature of grammar is no less arbitrary than that of vocabulary: usually in less than a couple of generations of formal schooling people who used to speak and still occasionally speak different dialects come to use the grammar imposed by the capital wherever needed. Finally, the fact that any human infant can be socialized into any language, and any adult without severe neurological impairments can learn any second language, also attests to the rather non-mysterious nature of grammatical diversity.

Humans Evolve

The new horizon opened by discursive metacognition changes the nature of evolution and of freedom. Those changes impact each other in several ways which pertain to individuation, need, world, care, and plasticity, while human freedom touches each of these areas and is impacted by all of them.

There is a persistent intellectual illusion that humans changed little since the species *Homo sapiens* acquired its biological makeup. Surely, we became more powerful due to technology, more knowledgeable thanks to scientific research, and mightier than Zeus in terms of our ability to deceive others and destroy things. However, the illusion would maintain, the core motives, types of social relations, etc., remain the same. In his international bestseller *Sapiens*, Yuval Harari writes:

> On a hike in East Africa 2 million years ago, you might well have encountered a familiar cast of human characters: anxious mothers cuddling their babies and clutches of carefree children playing in the mud; temperamental youths chafing against the dictates of society and weary elders who just wanted to be left in peace; chest-thumping machos trying to impress the local beauty and wise old matriarchs who had already seen it all. These archaic humans loved, played, formed close friendships and competed for status and power [. . .][237]

We have, however, zero evidence and no reason to believe that our ancestors were motivated and behaved so similarly to us. Harari here demonstrates the attitude that, modeling the term on anthropomorphism, can be called *modernomorphism*, seeing ancient humans through the paradigm we have for our contemporaries. This is deeply misguided. The way we are today, the way we carry out or suppress the general motives we are born with, is almost entirely a product of the world we built, and this is but a point on the continuum of human evolution. There is no guarantee whatsoever that teenage rebellions were present in the olden days – in fact, their presence in societies where females get pregnant at the earliest possible age is highly questionable. "Impressing local beauty" can have radically different meaning for all parties involved. Love is something that can be perceived in the context of ethical obligations or completely outside the realm of ethics. Yet we do not have to travel that far back in time to find out that "encountering a familiar cast of characters" might be an illusion. Many humans today feel utter disgust at the very mention of the types of relationships that many others consider acceptable, and lots of people would understand the meaning of consent very differently from what is understood by others. The way of treating each other, the conceptions of borders and property, the notion of ethnicity and the idea of religion – all those have profound, identity-forming effect on humans.[238] And all those are our own creation. The world we live in is our own creation, and it is a dynamic one, one that we re-create and alter all the time using the materials at hand, physical as well as conceptual.

Metacognitive Needs and World-Building

Like all animals, humans are born with a need for nutrition and with emotional needs. Yet, differently from any other animals except, perhaps, our primate brethren, we lack instincts, inherited multi-step behavioral patterns that are triggered by specific types of stimuli and provide ways of coping with environmental demands.[239] Human is "*Mängelwesen*, a deficient being no longer secured by instincts."[240] Yet not only food and shelter would be inaccessible for a newborn human being left to his own devices. Discursive metacognition, with which human beings are endowed, is, among other things, the ability to question mental contents of any kind. Questions like "Why?" and "How?" regarding links between different images, for example, are inevitable for an organism equipped with discursive metacognition, those that survived long enough to acquire language and thus instantiate their metacognitive capacity.

Why would asking questions about mental contents constitute a need? Because in a being capable of choosing between alternatives this constitutes a factor in formulating criteria for choice. In non-verbal animals, we can assume that the pleasure derived from this or that feeling, be it satisfying a sexual urge, relieving hunger, or getting their fill of affection are criteria for choice that are not questioned by the animal. In discursively intelligent animals, though, the ability to go meta on any mental content, not only more or less generalized images, to question the criteria themselves, is a factor in choice. This factor can become decisive, as humans can decide to formulate other choice criteria and give preference to different ways of deciding. For a being bereft of instincts and with behavioral options above and beyond those related to inherited general tendencies, questions begotten by discursive metacognition become a need to satisfy – a need to make sense of the world. This need, unlike the more specific needs like nutrition or affection, is both open and pervasive. It is tied not to a specific physiological function but rather to a cognitive capacity that can be used in a variety of ways, where the extent of this variety is virtually unlimited. Discursive metacognition also rules over all other needs: it can decide about the degree of their importance and thus guide human action in satisfying them. It is possible to observe a dog who gives preference to affection over food,[241] yet it is impossible to think of a dog who decides that some tasty food should be shied away from because of its high sugary content, or a dog that maintains that affection is an atavism that ought to be suppressed – this would require discursive metacognition, thinking about feeding and affection in terms of some non-sensory criteria external to them. For people, however, both feats are completely achievable: we fast due to dietary concerns and can try to suppress affection toward those we see as unworthy. We are free to do so – these are alternatives we can conceive of thanks to the language-enabled metacognition. This is also an inevitable sort of choices to make, as we have to make sense of the world. The evolution of humanity can be seen as developing ways of making sense of the world by building a world.

The Umwelt of humans is not limited to what is defined by the senses with which we are born. Some non-verbal animals can expand their world by using tools and developing socially transmitted habits, yet this expansion is limited in scope and determined in kind by the needs they inherit: the ape combines a longer stick from two shorter ones because it wants a banana, while humans put together microscopes to look for objects that cannot be seen with a naked eye, or an altar to invisible beings considered divine. The last two examples demonstrate how discursive metacognition not merely pushes the boundary of the Umwelt but constructs it. To think about entities too small for the eye to see requires a conception of size that is abstracted from imagery in such a way that it extends beyond the sense that acquired the content over which the scope of the sensed size ranges. To think of divine beings and their relation to humans is adding to the world a dimension that cannot be accessed by senses at all.

The world of the human being is thus a universe characterized by the "extendibility of everything that he relates to,"[242] ever expanding in the directions chosen by the human mind, as far as the human mind chooses to expand. This pertains to both Uexküll's perception and effect worlds. Our perception world extends as far as the instruments we devise can get – here Uexküll's own example of the cholera bacilli in water is very apt, yet not in the way he saw it.[243] Apes cannot see microbes, and thus they are not part of their world as observable objects. We cannot see them with the naked eyes, yet we can conceive of things smaller than what the eye can see, and thus choose to build devices that assist vision. We can also develop theories about what causes disease, examine correlations, etc.

The human universe expands even farther. We can think of relations that by their nature cannot be examined by the senses. Magical relations are of such quality – lack of susceptibility to sensation is their main feature. Religious faith frequently postulates relations and beings that cannot be examined sensorily – this is, in fact, the essence of the idea of faith as distinct from common belief. Finally, mathematics suggests the types of relations that are beyond our senses. Ideal triangles might have some approximations in the physical world, by the square of the square root of -2 is beyond what we can ever sense. All these are part of our world, we think of them and even grade each other on knowing or not knowing about them. Some impact our behavior quite strongly through commandments, the way people who have certain opinions about them behave, and more.

There can be facts we cannot conceive of[244] – some facts of which we cannot conceive of yet, some facts which we probably will never be able to grasp. The first category has multiple historical examples, from print to quantum effects. The second category, naturally, has no examples. However, the very possibility that we can consider their existence suggests that they may not be entirely out of reach. We might never be able to share the being of a bat, to have its phenomenological experience. Yet we can try to get closer, we can conceive of the limitations and possibilities of the bat's world, we can establish that bats have a life form different from ours in some respects and similar in other respects.[245] This gives us a distinctive freedom which is more than a mere choice between available alternatives. Not only do we have a much wider array of alternatives to consider than a non-verbal consciousness allows, but we also create radically new ones – not by recombining images but by going meta on whatever mental contents we have. We can ask questions like "Do we have enough information to make a decision?" – and thus go and look for more information, something that non-verbal animals cannot conceive of as their consciousness is limited to sensory contents and cannot ask questions about these contents' quality and quantity. We can question the paradigms through which we look at the world, and thus try and build new ones. This sort of questions can range from pragmatic specifics to questioning ways of treating the world around us, e.g., conceiving of empirical, spiritual, rational, and other ways of looking at the world. We can also question the quality of our ways of making decisions, thus giving rise to ethical thinking and changing our behavior according to the conclusions we reach.[246]

The Human Universe: Myths and Culture

Our discursive life form, then, is world-forming (*weltbildenden Dasain*)[247] – and this is the way of life enabled by discursive metacognition and an answer for the need of making sense of the world which is the outcome of having discursive metacognition. To satisfy this need, we need narratives, linguistically formulated explanations of possible mental contents that can be accessed through discursive metacognition. Narratives are usually provided by culture, a "guiding edifice" that would be functionally close to the role of instincts.[248] Historically, the first stopgap measure to provide a narrational answer to the "why" and "how" questions are myths.

Myths are explanatory narratives that treat of a variety of subjects: the beginnings of the universe, the origin of humankind, the reasons for ethical commands and prohibitions, and more. Usually they concern "profoundly important patterns, patterns that are very influential but too large, too deep and too imperfectly known to be expressed literally"[249] – myths use widely accessible terms and verbal imagery to address subjects that seem to be out of reach in terms of empirical accessibility and the available vocabulary. Myths also do not require verbatim transmission – the same meaning can be related in different words and even in different languages while still serving the purpose of providing an explanation for this or that phenomenon or justifying this or that behavior. Anthropologists frequently find similar stories in different human groups. For example, Vyacheslav Ivanov found a version of the Native American myth about the eagle nest ravager to exist among the ket people of Eastern Siberia and in the Sumerian literature.[250]

The existence of myths points to the main difference between human culture and the transmission of social habits in animals. Japanese macaques do not need linguistic narratives to transmit the habit of washing sweet potatoes, because what they are transmitting is not a justification of a certain practice but a habit, way of doing things. People, on the other hand, question the ways of doing things. To answer these questions, or to justify human practices, there is a need to address ways of doing things in terms of their relation to non-sensory criteria: justification, "we do X because of Y," is normative, not image-based. To address such a relation, one needs language with its apparatus of signs that can pertain to anything, not only to generalized images. This important distinction between transmitting a habit and providing justification for a common practice is frequently overlooked by those who analyze the phenomenon of culture.

Some researchers tend to treat culture as any transfer of behavioral patterns through learning, as a result of which different groups of the same species behave differently.[251] This definition, however, is very broad and would put under the same umbrella phenomena as diverse as classical conditioning and impressionist painting, phenomena that have very different characteristics in terms of motivations, technique, and, most importantly, the underlying cognitive capacities of the organisms that engage with them. More common among researchers is basing the definition of culture on learned information, e.g., conceiving of it as "*information capable of affecting individuals' behavior that they acquire from other members of their species by teaching, imitation, and other forms of social transmission.*"[252] Information here refers to anything that is acquired through learning and affects behavior. Even those who emphasize the difference between the information itself and its effect, conceiving of culture as a result of the interaction of humans with the information received through learning,[253] still see information as central to culture. However, these conceptions of culture also fail to provide an adequate account for the differences between dolphin calls and political economy, no less principally distinct than the terms in the previous example.

Culture deals with the transfer of meaning. All kinds of meaning that are transmitted by living beings for our purposes can be divided into three groups: data, information, and wisdom. The data are mere reflections of something that can be transmitted to another organism.[254] For example, the DNA carries genetic data encoded in the form of a sequence of paired nucleobases. In the right context, it can be interpreted and thus turned into a set of instructions for developing a new organism – and thus become information, that is, actionable data, data that affect and effect behavior. Everything that can be acted upon is data, which makes it the most general category; whatever is actually being acted upon is information. This equally applies to vervet monkey calls, lists of numbers that reflect weights, a link between an event and a behavioral response to it, etc. Information can be produced, transmitted, and interpreted by non-verbal animals, intentionally or unintentionally. Conveying the pattern of avoiding a certain object to the next generation by primates[255] is information transmission. Yet human culture, and specifically myths, have another crucial element: they suggest explanations, reasons for things to be this way and not otherwise, justifications of social practices; they answer the question "why?" – a metacognitive question. What myths transmit is not mere data that can be interpreted by its recipients and thus constitutes information: it attempts to satisfy their metacognitive needs, to tell them of reasons for certain ways of life, e.g., ethical commands, causes of various phenomena, the change of seasons, etc. This third, metacognitive aspect of meaning can be referred to as wisdom – something that relates to information to explain or evaluate it.

Here freedom enters as a crucial aspect of human evolution. The very *raison d'être* of myths is to try and influence the choice of those living beings who are endowed with discursive metacognition. It is because humans have choice that can rely on conscious selection of criteria for choosing rather than on comparing available options based on criteria that are not critically examined, that there is a need to explain and justify a certain way of life.

Traditional as modern human culture is, of course, wider than myth. It includes a whole variety of behavioral patterns transmitted from one generation to another: technology, ritual practices, ways of child rearing, etc. All these are infused with discursive metacognition – about all of them humans can ask the question "why?" and try to answer it. It is up to human individuals to evaluate and then accept or reject any answer to this question, within the limits set by the internalized cultural practices as they interact with the capacity for self-determined choice. In one cultural environment individuals will be inculcated into certain ways of behaving and punished not only for acting differently but also for questioning the wisdom of the received ways, so there we probably cannot expect widespread dissent – especially if the environment changes little, and the cultural ways are aligned with its demands. Another environment might be more permissive or even encourage critical thinking, so there we would expect more leeway for discursive metacognition to be expressed. In yet another situation, a society that imposed harsh strictures on questioning its mores is faced with radical challenges, e.g., epidemics, wars, or drought. The inadequacy of its customs, at least some of them, to what it encounters becomes evident to discursive beings who are capable to go meta on their sensations and thoughts. The discursive metacognition awakens to evaluate the state of affairs; and, perhaps, together with free choice, to offer different solutions. No animal society stays unchallenged for a long time, and human society experiences challenges most frequently thanks to the freedom of human beings who move from one region to another, thus spreading ideas as well as disease; create new tools and change the way they interact with flora and fauna, thus potentially stressing the natural environment to the degree it impacts the viability of the old ways of life; attack each other; and more. As a result, cultures are most likely to experience frequent assaults of discursive metacognition. This establishes freedom as a factor in cultural fitness, thus leading to the discussion on the possible evolution of culture. Yet before we proceed to the evolution of culture as driven by freedom, there is a need to address human individuation.

The Human Individual: Shaping Culture and Shaped By It

Human culture cannot but be created by human individuals: there is no other source for its components, be it myths or technology. This is why Hegel considers culture to be a self-alienated spirit, and the process of acculturation – the process of the self-alienation of spirit.[256] Cultural contents are created by humans, yet then they are externalized: formulated in language and become independent of their origin, carried forward by those who remember and relay them or recorded in some form. This way the cultural world, or, in out terms, the human universe, becomes an external reality to human consciousness – children internalize culture step-by-step in the process of acculturation (*Bildung*), part of the socialization. And yet at the same time it is the creation of human consciousness: these are humans who create, transmit, and change the elements of any culture.

On a superficial examination, it might seem that human children are molded by their cultural environment. Whatever general tendencies we inherit are useless without a supportive response from infants' caretakers: human children are born helpless and would not survive without them. A sucking reflex will bring no benefit without a caretaker providing the newborn with milk as is the predisposition to acquire language – without a proper linguistic environment. While we have no instincts to imprint, the significant others in our lives are those who always carry certain attitudes, demonstrate certain behaviors, inculcate us into particular ways of doing things. These are the mental materials with which we have to work. The language we learn is the one with which we not only communicate but also structure our world, cognize things and events. The vocabulary of this language is likely to shape important aspects of our way of thinking about the world. Lacking proper terminology might severely impair our ability to confront certain situations critically. It is hard to imagine science without the notion of the empirical as opposed to the hypothetical, for example. Similarly, certain concepts we are exposed to early on through discourse are likely to shape the way we conceive of events and act. A classic example here is the logic of hierarchy: the assumption that all phenomena form an order with the top and the bottom being different on some important scale, where the top is more preferable than the bottom. Inculcated into this paradigm, implicitly or explicitly, people apply it to whatever events and entities they encounter, to all meaningful content that comes their way.[257] When imposed on the relations between living beings, this inevitably leads to differences in the ethical treatment of different forms of life, and within human society – different individuals.[258] Yet the influence of society goes farther than equipping human individuals with the way to conceive of the world and has direct relevance to human freedom.

Agamben, standing on the shoulders of the long philosophical tradition going back to the Renaissance, notes that "man is the animal that must recognize itself as human to be human."[259] Born naked in terms of the way they see not only the world but also themselves, human individuals, as endowed with discursive metacognition, are entities that must conceive of all things including themselves and their own way of being. Here again human culture provides an answer. In all cultures people get names and are attached to their names so much that stopping to refer to an individual by his name causes major psychological upheavals that are routinely exploited by brainwashing techniques and other attempts to effect profound change in people's perception of themselves. Each person is socialized into a variety of roles – a member of a family as opposed to those who are not members of the same family; part of a clan, tribe, or nation as opposed to those who are not; bearer of a certain religion or ideology; carrier of certain rights presented as inalienable or privileges perceived as distinguishing; etc. These are parts of our identity even if we want to get rid of them: if the latter is the case, they define our struggles. Recognizing ourselves to be human in each individual case is having quite a bit of mental content that the said individual shares with some people and does not share with others. We do not just know ourselves, we always know ourselves *as something* – Heidegger's *as'-structure* is the condition of the possibility of human thinking, as it is the essence of discursive metacognition.[260] Culture provides the materials from which we construct this something we see ourselves as, the content for defining ourselves. And all this content, all those aspects of our identity are man-made fiction.

Here freedom enters again. While a specific cultural environment is a given for everyone, each individual has the capability to question and potentially change it, thanks to the power of choice and discursive metacognition, as it was noted in the last section. It is freedom enabled by discursive metacognition that enables people to overcome the alienation of individual consciousness and culture by understanding that culture is a creation of human spirit, and thus can be changed by human spirit.[261] This might start with individuation, where a person recognizes that she can be a molder of herself, as Picco della Mirandola suggested.[262] In fact, she is always a molder of herself – "the spirit makes itself what it is,"[263] whether by uncritically accepting society's suggestions or by trying to change the fare served by socialization. In the most basic sense, we are free to mold ourselves. Human freedom, aware of itself as free, is in this sense the way of overcoming the alienation between the individual and the world – something of which animal choice falls short.

Human freedom, as any freedom, is limited by circumstances, as noted earlier. A person burdened with immediate survival needs would have little mental resources to devote to questioning the merits of analytic philosophy. Another, raised in an environment that did not appreciate critical thinking, would need to make an extra effort to question the premises of the received view of the world and his own place in it. This way, the tension between the ability to choose and the available choices becomes the playground of human freedom, a very special one: discursive metacognition enables individuals to re-define the field of choice and conceive of radically new alternatives to choose from. Since we are endowed with language that, differently from the signaling systems of non-verbal animals, can communicate not only image associations but also new ideas to other humans, this freedom becomes the driving force of human evolution.

The Distinct Nature of Human Evolution

Human beings are embodied entities. The human life form is rooted in human physiology, it is enabled by it and depends on it – no culture would survive if it ignores the most basic needs of human beings, as its adherents will simply perish. Therefore, those researching culture in the context of evolution see cultural patterns as adaptive and as subjected to the selection pressures.[264] Those cultural contents that help with adapting to the environment, thus enabling their bearers to survive and pass their genes and culture on, will persist, while those that do not – will perish, frequently together with those who stick to them. The immediate problem with this approach is the existence and, in fact, the pervasive presence of cultural maladaptations – customs and related practices that by any count do not contribute to survival and thriving of humanity. These are quite widespread: resource-consuming habits that at times drive those who have them to the brink of extinction and beyond, extreme propensity to kill conspecifics that is not present to even a remotely similar degree in any other species, massive consumption of food that is known to be detrimental to health, culturally sanctioned use of intoxicants the adverse effects thereof on health are common knowledge, pollution of the environment that is known to be harmful, and more. While there are numerous attempts to justify such practices, their evolutionary disadvantage becomes clear if we take into consideration that those who engage in them know full well about their dangers, immediate and – in many cases – long term, yet persist nevertheless.

This last point is the reason why the attempts to explain persistent maladaptations by statistical models that calculate ratios and probabilities of mimicking existing behaviors versus behavioral innovation,[265] or in other similar ways that might be useful when applied to non-verbal animal societies, are inadequate wherever human evolution is concerned. Even with non-verbal animals this kind of approach is problematic: in animals capable of choice and sporting a cognitive structure sophisticated enough to have multiple generalized mental images relate to each other, social inheritance of behaviors might be too complicated to fit into the models built with genetic inheritance in mind. With humans, there is an added component – discursive metacognition. It redefines the world of individual preferences according to the criteria of its own making. Humans build their own universe from natural materials and processes, yet they arrange them according to the ways of their making, mostly frequently without regard to sensory association encountered in the past – and these may or may not be conducive to survival; or conducive to survival in different ways. They can be inherited socially, yet there is no telling how they will change. Most importantly, the choices human individuals make may impact the future in ways the significance of which, with the development of technology, exceeds that of non-human factors. This way freedom becomes the defining factor of not just human history but human evolution – a factor that is transformative rather than determinative, it changes the way human evolution proceeds rather than impacting it alongside other factors.

A few examples will help to illustrate the point. Disgust is a common emotion, it is hard to imagine a society where it would not play a role. It well might be that the neurophysiological disposition in which the feeling of disgust is rooted had serious evolutionary significance: it helped in avoiding dangerous foods, e.g., animal refuse and rotting corpses, and thus saved those who possessed it from dying of infectious diseases. However, this very physical aspect, once subjected to the working of discursive metacognition, developed into areas that have little to do with illness. People frequently feel disgust when grasping the meaning of certain signs, e.g., reading a story about a practice that is strongly rejected by social mores or by their own preferences that might be different from those common in their society. The notion of purity as opposed to contamination that appears in many religions as well as in secular thinking is also the notion of disgust stretched from its humble and utilitarian evolutionary roots all the way to the level of moral scale: many would judge as immoral a deed that has nothing to do with justice or mercy but is outside of the farther boundaries of common practice and thus elicits a disgust-like rejection.[266]

Another example pertains to the ability of human societies, i.e., humans whose ideas became definitive of a certain culture, to restrict the behavior of individuals by indoctrinating them into a certain set of views. People can develop a notion of the world that would define society as a unit the boundaries of which are not to be breached, the rules of which are natural to humankind, and the laws of which are binding. The idea of an Eternal Nation, the only right religion, workplace as a family, social class defined by the ownership over the means of production as determining human value, racial purity and genetic determination of human worth, etc., approximate this situation to various degrees. Those who buy into these notions will lock themselves into a world of people's own making:[267] they will be unlikely to even consider transcending it, might lack the vocabulary needed to conceive of the world being different. The results for most intents and purposes may be similar to the way Heidegger's non-verbal animals are captivated, locked in their world by the inability to transcend it.

The human universe has direct relevance also to the physiological aspects of evolution. Certain inherited physiological traits can become more widespread due to preference of potential mates or disappear for the same reason, if the preference is strong enough. This has been suggested by Darwin under the heading of sexual selection, yet not firmly established by research. However, it is quite clear that the physiological makeup of the modern human, rather different from our primate ancestors in terms of the natural protections from the elements, e.g., thick bodily hair, can coexist with survival only thanks to a material culture that includes practices of making clothes, using fire, and building efficient shelter. Here too human freedom, discursive choice plays the central role: these are human choices, culturally enshrined as traditions, that define how we protect ourselves from inclement weather, what kind of medicine we develop and whether we develop medicine at all rather than letting the sick die, and more.

This way, culture and the rest of the human universe that people erect becomes our true environment. What the genetic makeup is for a lion, what the natural surrounding processed through social behavioral patterns is for a chimp, history and culture are for a human. This is our characteristic as a form of life, having "history, civilization as 'second universe'" – the universe of our species' own making.[268]

This dual nature of our life form, being-impacted by the given cultural world and being-active in shaping and re-shaping it, is what defined the other characteristics of human evolution. Out needs are no longer limited to those dictated by the genetic endowment of *Homo Sapiens* – they are shaped by the universe we build. A newborn human needs food and affection; a teenager needs, in addition to that, very specific kinds of recognition by his peers, where the kinds of recognition are defined not only by the surrounding but also can be impacted by him. A person cultivated in a certain way would genuinely need reading and would suffer in the absence of it, while another can have, in addition, a desperate need for nicotine, which developed due to what he had decided to do earlier in life.

A human's relation to the world is one that, thanks to discursive metacognition, is disclosed to the human bearing the said relation as a such, as a relation – something that is not accessible to non-verbal animals since they cannot relate to non-sensory mental contents. The basic state of the human kind of being in the world, human Dasein, is anxiety (*Angst*) – the way human cognition relates to its own being-in-the-world, anxiousness about being thrown into the world which we sense, of which we think, in which we are to survive or perish, and in which we have various possibilities to pursue.[269]

This, in turn, also discloses human's freedom to this human's own self, Dasein as "*Being-possible,*" i.e., "*Being-free* for [*Freisein für*] the freedom of choosing itself and taking hold of itself."[270] This way, human care (*Sorge*), rooted in our thinking about our being in the world, *a priori* discursive metacognition, is the basis of willing, wishing, urge, addiction, etc., facets of life that in non-verbal animals are driven by desires and shaped to a rather small degree by uncritically accepted or rejected socially transmitted and personally acquired habits.

Finally, human plasticity is also deeply affected by freedom in the world. Firstly, our brains seem much more plastic than those of other animals. Summarizing many decades of neurological research, Costandi notes that "the structural and functional differences between individual brains probably outweigh their similarities [. . . the] brain is, to a large extent, unique, custom-built from the life experiences" starting with the embryonic period and continuing throughout lifetime.[271] This is the sort of brain that is needed for functioning in a world where one's survival depends on cultural information and wisdom, a brain that needs narratives to satisfy its metacognitive needs, a brain through which the organism guides its adjustment to circumstances the degree and direction of the variance of which is virtually unpredictable. Yet human plasticity if not confined to neuroplasticity. Our way of acting, our cultures are plastic too, as it has been noted multiple times earlier. We are also plastic in terms of storing and retrieving information. Discursive metacognition enabled us to think about remembering things, and from here the road to writing and then storing data on a variety of devices external to our brains is not too long. The same happens to processing capacity. For a non-human primate, having more capacious memory and quicker processing structures is helpful for survival. Humans can externalize big chunks of those: by writing books and using an abacus first and then proceeding to the Internet, capacious computers, and computerized neural networks that can externalize time- and resource-consuming processes like categorization and image recognition. Thus, the more advanced these technologies are, the less important are the abilities to remember more and calculate quicker – what's more important is having a more plastic brain that can focus on better choices.[272]

From what has been said it might seem that human evolution has no direction: it is shaped by individual choices that, justified by narratives, impact the world other humans are born into. There is no telling how and in what direction these will move. As Bookchin suggested, we chose hierarchy – yet we could have chosen something else.[273] Religion, technology, medicine, family structure, morality – all these could have been different, if they were to come to life at all. And yet I will try to argue that there is some directionality to human evolution and history.

Human Evolution: From Myth to Philosophy and Liberty

The combination of discursive metacognition and freedom makes it very likely for at least some of us to try and question the narratives we are given, including myths. The ability to ask "why?" together with the need to answer this question is the ability to ask other similar questions regarding the merits of this or that explanation. This is what Heidegger, following Aristotle, referred to collectively as an a*pophantic discourse* (λόγος ἀποφαντικός), discourse that "seeks to point out what is meant itself."[274] As such, this discourse is essentially revealing, or disconcealing (ἀ-ληθεύειν), the logos of truth and falsity where truth (ἀλήθεια) is exposing then nature of things to the inquiring cognition.[275] This ability (*Vermögen*) of human agents for metacognitive judgment, coupled with the freedom to engage in it,[276] entails the possibility (*Möglichkeit*) of normative evaluation of the received as well as self-generated mental content of any kind. And this carries us from myth to philosophy.

Myths do not care for truth and falsity. This is why they are bound to be lacking when questioned by human reason. This questioning that can address and does address not only external contents but itself is philosophy – it consists in this questioning, in philosophizing.[277] As such, it is the ultimate answer to the metacognitive needs – not a specific answer to a specific question, but a way of thinking that matches the need for seeking reasons. As Midgley notes, unexamined life is not only not worth living – "unexamined life is unlivable to man."[278] Metacognitive examination is the essence of our life form.

Philosophizing is also evolutionarily useful. We lack instincts, and our general tendencies do not form a coherent system that will enable us to survive – which makes us inherently confused beings. Yet "the cultures by which we try to make sense of those impulses often work very badly."[279] This is not only a conclusion at which one arrives after studying ancient and modern history – this is a necessity. The human universe, in all its aspects relevant to our life, changes much quicker than our biology – in fact, it changes on an entirely different time scale; the reasons have been explained earlier, and not the least of those is our freedom. No fixed tradition can provide the answers necessary to cope with emergent challenges. The only capacity we possess to help us survive in this world as a species is our ability to question things, put forward possible specific answers and more general ways of coping, and evaluate them critically. Various disciplines that spring up to provide such responses, from tool building and medicine to political organization and religious faith, would be no better than fixed traditions if they are not being constantly subjected to critical assessment – and this is what philosophy is concerned with. Philosophy is acting upon the realization that the human universe is a work of human freedom, and thus it can and needs to be critically examined.

A human society that tries to suppress this critical evaluation of our ways of life, when confronted with challenges for which its tradition provides no adequate response, is likely to fare worse than a society that encourages critical thinking. Mill noted that one of the benefits of political liberty is that different ideas are brought up for consideration, and this increases the chances of selecting the better ones.[280] It does not guarantee, of course, that this will be the case – yet it at least provides a chance. On the other hand, a society that does not encourage free exchange of ideas in the long term seems doomed to encounter a challenge for which it would not have a proper response.

The last consideration points out the vector of human evolution. In the long run, those groups that provide the political and cultural infrastructure for freedom to lead to the bettering of human condition are more likely to survive and flourish. That is, of course, if we do not use our freedom to extinguish human life on Earth prior to the firm establishment of the culture of freedom.

Conclusion and the Road Ahead

This book attempts to give a convincing account of the *how* of freedom-in-the-world, showing the ways freedom can be instantiated in material agents. It began by outlining the desiderata of freedom and specifying the necessary characteristics of a free entity. Then it addressed the problem of freedom, one that has been keeping philosophers busy for a long time – the existence of free choice in a determined physical world. Examining possible solutions to this problem, it indicated that the only feasible conception of freedom is that of the compatibilists who are trying to reconcile choice and determinism. Insofar as the compatibilist conception of freedom is the only conception of freedom that can possibly exist in the material world, the challenge is to develop an account of how such freedom is instantiated in the world.

I argued that life answers the call – living entities have a self that is functionally separate from its environment, a self that determines to varying degrees its own development. Since life evolves, the most reasonable place to start looking for freedom as self-determined choice is living things. Examining different forms of life, I concluded that animals hold the potential for self-determined choice, and that the potential is realized for the first time in animals that possess representing consciousness. A self that can compare retained images with intuitions can behave this way or that and be a self that makes choices.

The choice of non-verbal animals is limited since they lack discursive meta-cognition, the ability to consider their own cognitions and relate to their own mental content. This does not allow them to envision radically new alternatives, those that cannot be conceived of by generalizing over retained sensory images and re-combining them. This ability is acquired with the development of discursive metacognition, the capacity to relate to any mental content, not only to sensory imagery but also to the relations between images, the relations between these relations, etc. This ability is characteristic of humans and it is carried out using language: labelling mental contents and operating on them by means of words and grammatical relations. Metacognition, through discursive intelligence, opens a new horizon of freedom, enabling a virtually unlimited field of choice that includes culture, normativity, science, and more. This makes human freedom break out of the boundaries of animal choice and consciously create its own world.

I also argued that freedom evolves together with biological evolution and at a certain point starts altering the latter's course. The precursors of freedom, namely the organism's ability to control itself, perception, and memory, evolve with the increased sophistication of animals. This sophistication, I argued, is something that ought to be expected: given that genetic mutations can lead to both simplifying the structures underlying behavior and making them more complex, the road to greater sophistication is always open. Each more sophisticated type of organism alters the structure of the pressures of natural selection: once a certain more advanced ability, e.g., memory, appears, it becomes advantageous at least for some animals, and thus those of their offspring that possess it have a higher chance of survival. Because of that it becomes more probable that the ability will evolve further over the coming generations. When the evolution of animals gives rise to representing consciousness, self-determined choice between alternatives becomes possible. At this point, natural selection starts favoring the ability to choose better: taking more alternatives into consideration, re-combining mental images more efficiently, etc. Genetically inherited behavioral patterns, or instincts, now become increasingly less important in comparison to the abilities that underlie better choice.

The increased sophistication of the nervous system, with its ability to consider more and more mental content, at a certain point crosses a critical threshold – it becomes able to address non-sensory mental content, leading to the development of meta-cognition and language. This, in turn, leads to the development of culture and other normative domains, as well as philosophy. The latter is the epitome of discursive metacognition, as it is not limited to any specific area of consideration and can critically examine its own foundations. It also impacts the course of evolution, as now humans can think of altering the course of their own development and effectively do that.

This analysis of freedom and its evolutionary development might become a starting point for several directions of philosophical research, some of which I will try to suggest now.

FREEDOM AND HUMANITY, HUMAN GOOD, AND ETHICS

Foucault notes that from the moment philosophy became anthropology, it radically altered our conception of the animal world.[1] At this very moment it also becomes self-creative: philosophical anthropology, whether disciplined and aware of its role, or still semi-mythical, not merely describes humans' place in nature but defines it. This is because discursive freedom enables us to alter the world – not merely this or that aspect of our surrounding, but radically. This includes designing ourselves.

Freedom invalidates the reduction of humanity to any set of pre-determined principles.[2] The evolution of non-verbal animals endowed with choice made instincts less important for survival than neural structures that enable better choosing. With the appearance of discursive intelligence, we lost inherited behavioral patterns that are triggered by specific stimuli. This left us with vague urges, general tendencies, the satisfaction of which requires acquisition of a variety of ways to act that are usually acquired through socialization. As a species, we have no choice but to define ourselves.[3] As individuals, we are capable of changing what we are. This alters the philosophical nature of the question of what we essentially are: the answer becomes categorically different from the one given in the case of other entities. It can no more refer to some fixed traits or contents but becomes synonymous with the main thing we are able to do.

As *homo sapiens* is a discursively free being, the only thing we can say about the essence of human nature is that we are free, capable of defining ourselves. This is what sets us apart from inanimate nature and from the plant life form. This is what makes us both different from other animals that are capable of choice and similar to them. We share both material embodiment and the capacity to choose – and this similarity, as I will try to argue soon, needs to be explored for possible ethical implications. The point of difference, though, is that our freedom is different in that it can design ourselves in the way animals cannot – design conceptually, having a new kind of human in mind, new in the way it thinks, behaves, looks, procreates, etc.

This has direct implications for the understanding of the human good and ethics. If our essence is freedom, then it is impossible to state what the human good is as a set of pre-defined principles rooted in our specific biological nature as observed and conceptualized, deduced from the observations made of human behavior, or inferred from the way we tend to think. This is because we change ourselves, so any such principle would apply only conditionally. Happiness, pleasure, type of willing, even physical goods all become contingent rather than universal in the human-defined world that can be rebuilt based on free choice.

What kind of ethics can befit a discursively free agent? I would argue that such ethics needs to start with the formal essence of humanity – freedom. If we are to think about the broadest basis of human good, that would be the ability to exercise free choice. Such ability has two aspects: removing constraints and enabling better, i.e., genuinely self-determined choices. The principles of ethics then, both at the level of personal choice and at the macro level of right, should be constitutive of the reality of self-determination.

Removing constraints has been actively discussed since the dawn of philosophy. The enabling aspect of freedom, however, is attended much less. Yet just like with animal choice, there are certain factors that make choosing more effective and meaningful. With non-verbal animals, these would be the number of alternatives available and the ability to process data describing these alternatives. For our animal brethren, the scope of choice is limited by the capacity of their neural structures and the range of available alternatives. Human neural structures enable discursive meta-cognition, and with it – a virtually unlimited scope of choice in principle. The scope of choice in practice, though, is determined by the breadth of knowledge and the means that are available for formulating and analyzing alternatives, as well as a safe socio-cultural space to try different ways of doing things. This has specific implications in terms of access to education and the type of education, access to resources that enable experimentation with the way of life, and more[4] - the implications that might be worth exploring.

ANIMAL RIGHTS

If we are to consider ethically the ability to choose, then we need to take a new look at animal rights. Traditionally, the question of animal rights has been tied to sentience: sentient beings have interests, they feel pain and experience pleasure, and thus, if we are to consider these as a basis for the ethical treatment of human beings, we cannot ignore them in non-verbal animals. This way of arguing, however, is not pursued in the discourse of human rights: it is hard to establish an entitlement to some sort of actions based on the ability to feel pain; even to argue for the avoidance of causing pain to those who can experience it, we need another basis which will ground the idea that causing pain to those who do not enjoy it is ethically inadmissible. Thus, the discourse of rights is usually based on autonomy, the ability to act independently in the world. "This is because to respect autonomy is to allow persons to form, revise, and pursue their own conceptions of the good. For example, there is a long tradition within political liberalism in which substantive liberal institutions such as freedom of expression, religious tolerance, and the freedom of association are primarily justified by appeal to the intrinsic value of autonomy"[5]

It is on this count that animals have been considered as lacking any relevance to the discourse of rights since at least the Enlightenment. The type of autonomy required for an agent to be considered a bearer of rights would be *moral autonomy* in the broad sense of individuals who are capable of making moral judgments about the rightness or wrongness of actions, which underlies the ability to recognize and honor rights, and have motives that can be judged morally. Since animals do not demonstrate moral judgment, they cannot be morally autonomous.[6]

It is due to the lack of this kind of autonomy that animals have had a rather bad historical career as far as rights are concerned. From ancient totemism, worship of animal spirits, identifying divinities with animals and animals with divinities, animals came to be considered a resource, de-animated to the status of the supply of labor, energy, and entertainment. This development culminated in "the present tragedy and farce that combine slaughterhouses, pet shops and global safaris,"[7] as well as circuses and biology classes where animal life is considered to be a suitable expense for enlarging the biological knowledge of high school students. Not surprisingly, this dramatic drop of animal status paralleled the ascent of the right of humans during the Enlightenment: it was the moral autonomy, imputed to humans and denied to animals, that drove the ascent of human rights and the depreciation of animals to the status of a resource.

In what follows I am not attempting to present a comprehensive account of how the capacity for choice can underlie non-verbal animals' rights. I am merely trying to point to a direction that can generate some questions that seem worthy of consideration and that, perhaps, can challenge the predominant way of thinking about the rights of non-human animals.

The reasoning behind the assignment of rights based on moral autonomy can be challenged. Autonomy at its basis is the capacity for behavioral independence. An autonomous self is a self that can move itself, exercise a degree of independence in its behavior from the outside world. The discourse of autonomy is relevant to the entities that can choose how to behave and irrelevant to those that cannot. Therefore, discussing the implications of autonomy as related to rights is principally possible for those animals that can exercise choice.

Another problem with the arguments for moral autonomy as necessary for rights is that it does not distinguish between moral agents and moral patients, entities that can be held morally responsible and those who figure as ethically significant. Moral autonomy pertains to moral agents: it makes little sense to chastise dogs for stealing food with an expectation of moral correction, as they do not have the conceptual apparatus necessary to understand the idea of theft. Instead, behavioral correction to prevent similar actions in the future might be in order, e.g., generating negative emotions related to the disapproval by beloved human figures; same applies to pre-verbal children. However, this does not mean that dogs cannot be moral patients, subjects to moral treatment by those who have the conceptual capacity to behave morally, i.e., humans. A similar analysis applies to the assignment of rights based on the ability to recognize and honor rights. And entity that can recognize rights can also be held responsible for violating the rights of others, yet being unable to play the role of an agent of rights might not necessarily mean that one cannot be a bearer of rights. When we consider humans who, due to early age or a cognitive impairment, cannot exercise rights that require conceptual thinking, like freedom of speech or conscience, we still assign them some basic rights, e.g., freedom from undue restrictions on movement or the right to life, their inability to recognize and honor the rights of others notwithstanding. Perhaps, we can consider the rights of non-verbal animals in a similar way. An entity that can choose is autonomous without being endowed with discursive intelligence and thus capable of rational agency, and if autonomy deserves ethical respect and accommodation, then all autonomous entities can be considered in this context.

From here, viewing autonomy, self-determined choice that can be non-discursive, as a basis of rights can also outline the possible limits for animal rights, just like it limits the rights of humans who cannot exercise conceptual reasoning. If it is the capacity to choose that is important for defining rights, then the limits of this capacity limit their scope. The rights the exercise of which depends on having discursive intelligence are irrelevant to non-verbal agents: it makes no sense to provide a right to vote or to worship to those who cannot understand the meaning of elections and religious belief. On the other hand, the right to life and freedom from arbitrary restraint can be exercised by non-verbal entities, and thus animals should be considered when discussing this sort of rights.

Much remains to address in regards to the line of reasoning outlined here. The distinction between the agent of rights and the bearer of rights makes sense on the natural right account, an account that sees the basis of right in the capacities of their bearer and argues that this capacity is enough for claiming the rights.[8] Yet how would it fare if we consider rights as contractual, based on the agreement between free individuals who understand the meaning of rights? How do capacities underlie rights, is it enough to have a capacity to claim a related right? If we grant rights to non-verbal animals, would discursive intelligence play a role in giving preference to the rights of discursively intelligent entities over the rights of those who lack this capacity when we consider rights that can be exercised by both? It seems, though, that this inquiry might be worthy of pursuing.

ARTIFICIAL FREEDOM?

I argued that life is a necessary condition for freedom in the world. However, with the recent developments in the field of artificial intelligence, the possibility of creating artificial freedom should be discussed. As it was the case with animal rights, I am not trying here to argue for a conclusion, merely to outline a possible subject if inquiry and suggest several questions.

It seems clear that computer programs, no matter how sophisticated, cannot be considered free. What they do is defined by the human programmers, so they do not determine themselves in any shape or form. The example of a target-seeking torpedo that Jonas discusses is instructive in this regard:[9] the "degree of freedom" in the movement it exercises and how it exercises it is fully determined by the engineers who built it; 'freedom' in the 'degree of freedom' is merely a metaphor.

Similarly, common neural networks cannot be considered free.[10] These are computer programs that consist of nodes (neurons, a computational metaphor of the neurons in the animal nervous system), each of which embodies a formula with a numeric coefficients that can be changed based on the result of the computation. When a neuron computes the result of a set of inputs, it is compared to the a number; if the result differs from the target number, the coefficient is changed, and the computation is repeated – until the results of the computation come as close as possible to the target. Such neurons can be connected into networks, where the output of one layer of neurons is the input for the next layer, or even for themselves; different configurations of neural networks can be built, e.g., by having different number of neurons in different layers, different patterns of linking neurons, etc.

Neural networks handle a variety of tasks. For example, a neural network can be built in order to categorize objects into moving in front of the camera and not moving in front of the camera. It then can be improved and categorize objects into those that move quickly or slowly enough in order to collide with the self-driving vehicle equipped with the neural network, and those that do not. The network will be supplied with a video stream, the neurons' formulas will output results which will be aggregated by the network into the final decision, and these decisions will be compared to the expected result. This comparison will provide the degree of accuracy of the identification of objects. Based on this identification, the network will change the coefficients for the formulas of each neuron and try again – until satisfactory results are achieved. Here the program is more sophisticated than Jonas's torpedo, yet it is by no means freer. The parameters, the structure, the starting values and the way they change – all these are defined by the programmers. The network is not determining itself in any way – it is not building itself and not making decisions outside the range and the method coded into it.

Things start getting more complicated when we think of neural networks that can create other neural networks. No more changing values of the formulas of its own nodes but creating new nodes and then new networks, where the formulas for the nodes, the way their output is aggregated, what type of data they use, what kind action the machine takes based on the output, etc., characteristics of a neural network are defined by another neural network. Then the output of the first can become the input of the second, and the output of the second – the input of the first. This way, one neural network goes meta on another in some sense – not exactly the sense in which human mind can consider its own thoughts as in normative evaluation of its own intentions, but in some other sense the meaning of which is yet to be clarified. It is unclear because this loop is not determined any longer by what has been coded into the neural network initially. No parameter or range of values is under the control of the software developer – the size of the network, the formula of the neurons, their quantity, etc. are now up to the network itself. In a sense, here the initial conditions are created for something that will keep defining itself as long as there is a power supply – and the control of power supply can be assigned to the network as well.

Several questions become relevant at this point. Would such network be self-relating? Would it determine itself and create itself, and if yes, in what sense? Finally, is there a sense in which it can be considered alive? Would its choices be determined by it? Is there a sense in which it is making independent decisions?

Much inquiry is required to try and answer these questions. Yet thinking about them, even if it will not affirm or reject the possibility of artificially created computational entities that can evolve into choosing selves, may still help us in clarifying the meaning of freedom.

This book suggests a conceptual and evolutionary understanding of choice that breaks with the traditional scope of debates on the possibility of free will. It attempts to add some conceptual flesh to the skeleton of the compatibilist approach. I hope that it establishes a firmer basis for further exploration of the meaning of freedom.

References

Adamo, S. A. (2016). Do insects feel pain? A question at the intersection of animal behaviour, philosophy and robotics. *Animal Behaviour, 118*, 75-79. doi: 10.1016/j.anbehav.2016.05.005

Addessi, E., Mancini, A., Crescimbene, L., Padoa-Schioppa, C., & Visalberghi, E. (2008). Preference Transitivity and Symbolic Representation in Capuchin Monkeys (Cebus apella). *PLoS One, 3*(6), e2414. doi: 10.1371/journal.pone.0002414

Adler, M. J., & Institute for Philosophical Research (Chicago Ill.). (1958/1973). *The idea of freedom.* Westport, Conn.: Greenwood Press.

Agamben, G. (2004). *The open: man and animal* (K. Attell, Trans.). Stanford: Stanford University Press.

Agamben, G. (2015). *The use of bodies* (A. Kotsko, Trans.). Stanford, California: Stanford University Press.

Albahari, M. (2006). *Analytical Buddhism : the two-tiered illusion of self.* Basingstoke; New York: Palgrave Macmillan.

Alcock, J. (2001). *Animal behavior: an evolutionary approach* (7th ed.). Sunderland, Mass.: Sinauer Publishers.

Alpi, A., Amrhein, N., Bertl, A., Blatt, M. R., Blumwald, E., Cervone, F., . . . Wagner, R. (2007). Plant neurobiology: no brain, no gain? *Trends in Plant Science, 12*(4), 135-136. doi: 10.1016/j.tplants.2007.03.002

Arendt, H. (1961). *Between past and future, six exercises in political thought.* New York: Viking Press.

Arendt, H. (1968). *Between past and future; eight exercises in political thought* (Enl. ed.). New York: Viking Press.

Aristotle. (1984). *The complete works of Aristotle: the revised Oxford translation* (J. Barnes Ed.). Princeton, N.J.: Princeton University Press.

Aristotle. (1986). *De anima (On the soul)* (H. Lawson-Tancred, Trans.). New York: Penguin Books.

Atance, C. M. (2008). Future Thinking in Young Children. *Current Directions in Psychological Science, 17*(4), 295-298.

Atkins, P. W. (1984). *The second law*. New York: Scientific American Library: distributed by W.H. Freeman.

Avargues-Weber, A., & Giurfa, M. (2015). Conceptual Learning by Miniature Brains. In E. Margolis & S. Laurence (Eds.), *The conceptual mind: new directions in the study of concepts* (pp. 3-27). Cambridge, Massachusetts: MIT Press.

Ayer, A. J. (1954/1982). Freedom and Necessity. In G. Watson (Ed.), *Free will* (pp. 15-23). New York: Oxford University Press.

Balaguer, M. (2010). *Free Will As an Open Scientific Problem*. Cambridge, Mass: A Bradford Book.

Bartal, I. B.-A., Decety, J., & Mason, P. (2011). Empathy and Pro-Social Behavior in Rats. *Science, 334*(6061), 1427-1430. doi: 10.1126/science.1210789

Barth, A. L. (2018). Tool Use Can Instantly Rewire the Brain. In D. J. Linden (Ed.), *Think Tank: Forty Neuroscientists Explore the Biological Roots of Human Experience* (pp. 60-65): Yale University Press.

Beauchamp, T. L. (2005). Who Deserves Autonomy, and Whose Autonomy Deserve Respect? In J. S. Taylor (Ed.), *Personal autonomy : new essays on personal autonomy and its role in contemporary moral philosophy* (pp. 310-329). New York: Cambridge University Press.

Bédécarrats, A., Chen, S., Pearce, K., Cai, D., & Glanzman, D. L. (2018). RNA from Trained Aplysia Can Induce an Epigenetic Engram for Long-Term Sensitization in Untrained Aplysia. *eNeuro, 5*(3).

Berlin, I. (2002). *Liberty: incorporating four essays on liberty* (H. Hardy Ed.). Oxford: Oxford University Press.

Berns, G. (2017). *What it's like to be a dog: and other adventures in animal neuroscience* (First edition. ed.). New York: Basic Books.

Bhagavad Gita: The Song Selestial. (1965). (E. Arnold, Trans.). New York: The Heritage Press.

Bigger Brains: Complex Brains for a Complex World. (2/8/2016). *Human Characteristics: Brains*. Retrieved 12/27/2017, from http://humanorigins.si.edu/human-characteristics/brains

Bischof, N. (1980). On the Phylogeny of Human Morality. In G. S. Stent (Ed.), *Morality as a biological phenomenon: the pre-suppositions of sociobiological research* (Rev. ed., pp. 48-66). Berkeley: University of California Press.

Bloeser, C., & Stahl, T. (2017). Hope. In E. N. Zalta (Ed.), *The Stanford Encyclopedia of Philosophy (Spring 2017 Edition)* (Spring 2017 ed.): Metaphysics Research Lab, Stanford University.

Bode, H., Berking, S., David, C. N., Gierer, A., Schaller, H., & Trenkner, E. (1973). Quantitative analysis of cell types during growth and morphogenesis in Hydra. *Wilhelm Roux' Archiv für Entwicklungsmechanik der Organismen, 171*(4), 269-285. doi: 10.1007/BF00577725

Boethius. (525/1999). *The consolation of philosophy* (V. E. Watts, Trans. Rev. ed.). London; New York: Penguin Books.

Bontempi, B., Demir, C. L., Destrade, C., & Jaffard, R. (1999). Time-dependent reorganization of brain circuitry underlying long-term memory storage. *Nature, 400*(6745), 671-675. doi: 10.1038/23270

Bookchin, M. (2005). *The ecology of freedom: the emergence and dissolution of hierarchy*. Oakland, CA: AK Press.

Boyd, R., & Richerson, P. J. (2005). *The origin and evolution of cultures*. Oxford; New York: Oxford University Press.

Brenner, E. D., Stahlberg, R., Mancuso, S., Vivanco, J., Baluška, F., & Van Volkenburgh, E. (2006). Plant neurobiology: an integrated view of plant signaling. *Trends in Plant Science, 11*(8), 413-419. doi: 10.1016/j.tplants.2006.06.009

Brosnan, S. F., & de Waal, F. B. M. (2003). Monkeys reject unequal pay. *Nature, 425*(6955), 297. doi: 10.1038/nature01963

Bugnyar, T., & Heinrich, B. (2006). Pilfering ravens, Corvus corax, adjust their behaviour to social context and identity of competitors. *Animal Cognition, 9*(4), 369-376. doi: 10.1007/s10071-006-0035-6

Cairns-Smith, A. G. (2009). An Approach to a Blueprint for a Primitive Organism. In C. H. Waddington (Ed.), *The Origin of Life: Toward a Theoretical Biology* (Vol. 1, pp. 57-66). New Brunswick, NJ: AldineTransaction.

Campbell, J. K. (2011). *Free will*. Cambridge; Malden, MA: Polity Press.

Cardona, G. (2014). Paninian Grammarians on Agency and Independence. In M. R. Dasti & E. F. Bryant (Eds.), *Free will, agency, and selfhood in Indian philosophy* (pp. 85-111). New York: Oxford University Press.

Chisholm, R. M. (1982). Human Freedom and the Self. In G. Watson (Ed.), *Free will* (pp. 24-35). New York: Oxford University Press.

Chomsky, N. (2007). Biolinguistic Explorations: Design, Development, Evolution. *International Journal of Philosophical Studies, 15*(1), 1-21. doi: 10.1080/09672550601143078

Chomsky, N. (2016). *What kind of creatures are we?* New York: Columbia University Press.

Church, R. M. (1959). Emotional reactions of rats to the pain of others. *Journal of Comparative and Physiological Psychology, 52*(2), 132-134. doi: 10.1037/h0043531

Cicero, M. T. (1928). *De re publica, De legibus* (C. W. Keyes, Trans.). London, New York: W. Heinemann; G. P. Putnam's sons.

Clayton, P. (2009). *In Quest of Freedom: The Emergence of Spirit in the Natural World*: Isd.

Coetzee, J. M. (2016). *The Lives of Animals*. Princeton: Princeton University Press.

Costandi, M. (2016). *Neuroplasticity*. Cambridge, MA: The MIT Press.

Cox, C. J., Foster, P. G., Hirt, R. P., Harris, S. R., & Embley, T. M. (2008). The archaebacterial origin of eukaryotes. *Proceedings of the National Academy of Sciences, 105*(51), 20356.

Darwin, C. (1859/2017). *On the Origin of Species*. Minneapolis: Race Point Publishing.

Darwin, C. (1872/2009). *The expression of the emotions in man and animals* (4th ed.). New York: Oxford University Press.

Darwin, E. (1794). *Zoonomia; or, The laws of organic life* (Vol. I). London: J. Hohnson.

Dasti, M. R. (2014). Nyaya's Self as Agent and Knower. In M. R. Dasti & E. F. Bryant (Eds.), *Free will, agency, and selfhood in Indian philosophy* (pp. 112-136). New York: Oxford University Press.

Davidson, D. (2004). *Problems of rationality*. New York: Clarendon Press; Oxford University Press.

de Chardin, P. T. (1955/2008). *The Phenomenon of Man* (J. Huxley, Trans.). New York: HarperCollins.

de Waal, F. B. M. (1999). Apes With an Oeuvre. *Chronicle of Higher Education, 46*(13), B6.

de Waal, F. B. M. (2016). *Are we smart enough to know how smart animals are?* (First edition. ed.). New York: W. W. Norton & Company.

de Waal, F. B. M., Macedo, S., Ober, J., & Wright, R. (2006). *Primates and philosophers: how morality evolved*. Princeton, N.J.: Princeton University Press.

Dennett, D. C. (1992/2014). The Self as a Center of Narrative Gravity. In F. S. Kessel, P. M. Cole, D. L. Johnson & M. D. Hakel (Eds.), *Self and Consciousness: Multiple Perspectives* (pp. 103-115): Taylor & Francis.

Dennett, D. C. (1995). *Darwin's dangerous idea: evolution and the meanings of life*. New York: Simon & Schuster.

Dennett, D. C. (2003). *Freedom evolves*. New York: Viking.

Dennett, D. C. (2015). *Elbow room: the varieties of free will worth wanting* (New edition. ed.). Cambridge, Massachusetts; London, England: MIT Press.

Dennett, D. C. (2018). *From Bacteria to Bach and Back: The Evolution of Minds*: W. W. Norton, Incorporated.

Descartes, R. (1642/1996). *Meditations on first philosophy: with selections from the Objections and Replies* (J. Cottingham, Trans. J. Cottingham Ed. Rev. ed.). New York: Cambridge University Press.

Descartes, R. (1989). *The passions of the soul* (S. Voss, Trans.). Indianapolis: Hackett Pub. Co.

The Dhammapada. (2010). (V. J. Roebuck, Trans.): Penguin Adult.

Diels, H. (1903). *Die fragmente der Vorsokratiker griechisch und deutsch*: Weidmannsche buchhandlung.

Distin, K. (2011). *Cultural evolution*. Cambridge: Cambridge University Press.

Dobzhansky, T. (1955). *Evolution, genetics, and man*. New York: Wiley.

Dobzhansky, T. (1956). *The biological basis of human freedom*. New York: Columbia University Press.

Dobzhansky, T. (1964). Biology, Molecular and Organismic. *American Zoologist, 4*(4), 443-452.

Doolittle, W. F. (1999). Phylogenetic Classification and the Universal Tree. *Science, 284*(5423), 2124-2128.

Dunn, J. (2018). Why Can't Apes Talk? *Sapiens*. Retrieved October 27, 2019, from https://www.sapiens.org/language/primate-speech/

Dupeyroux, J., Viollet, S., & Serres, J. (2019). Polarized skylight-based heading measurements: A bio-inspired approach. *Journal of The Royal Society Interface, 16*, 20180878. doi: 10.1098/rsif.2018.0878

Eagleton, T. (2000). *The idea of culture*. Malden, MA: Blackwell.

Eisen, A., & Konchok, Y. (2018). *The enlightened gene: biology, Buddhism, and the convergence that explains the world*. Lebanon: ForeEdge.

Elwood, R. W. (2011). Pain and Suffering in Invertebrates? *Institute of Laboratory Animal Resources Journal, 52*(2), 175-184.

Fantl, J. (2017). Knowledge How. In E. N. Zalta (Ed.), *The Stanford Encyclopedia of Philosophy* (Fall 2017 ed.).

Firn, R. (2004). Plant Intelligence: an Alternative Point of View. *Annals of Botany, 93*, 345-351.

Fischer, J. M. (2007). Compatibilism *Four views on free will* (pp. 44-84). Malden, MA: Blackwell Pub.

Fischer, J. M., Kane, R., Pereboom, D., & Vargas, M. (2007). *Four views on free will*. Malden, MA: Blackwell Pub.

Fitch, W. T., de Boer, B., Mathur, N., & Ghazanfar, A. A. (2016). Monkey vocal tracts are speech-ready. *Science Advances, 2*(12). doi: 10.1126/sciadv.1600723

Flombaum, J. I., & Santos, L. R. (2005). Rhesus Monkeys Attribute Perceptions to Others. *Current Biology, 15*(5), 447-452. doi: 10.1016/j.cub.2004.12.076

Floridi, L. (2010). *Information : a very short introduction*. Oxford: Oxford University Press.

Foucault, M. (2006). *History of madness* (J. Murphy & J. Khalfa, Trans. J. Khalfa Ed.). New York: Routledge.

Frankfurt, H. G. (1969). Alternate Possibilities and Moral Responsibility. *The Journal of Philosophy, 66*(23), 829-839. doi: 10.2307/2023833

Garfield, J. L. (2014). Just Another Word for Nothing Left to Lose. In M. R. Dasti & E. F. Bryant (Eds.), *Free will, agency, and selfhood in Indian philosophy* (pp. 165-185). New York: Oxford University Press.

Glossary of Technical Terms. (2013) *Diagnostic and Statistical Manual of Mental Disorders*: American Psychiatric Association.

Godfrey-Smith, P. (2007). Conditions for Evolution by Natural Selection. *Journal of Philosophy, 104*(10), 489-516.

Godfrey-Smith, P. (2016). *Other minds: the octopus, the sea, and the deep origins of consciousness* (First edition. ed.). New York: Farrar, Straus and Giroux.

Goldenfeld, N., & Woese, C. (2007). Biology's next revolution. *Nature, 445*(7126), 369-369.

Goodman, W. (1987, July 6, 1987). I. B. Singer Profiled on PBS, *The New York Times*, p. 1.

Graupe, D. (2013). *Principles Of Artificial Neural Networks (3rd Edition)* (Vol. 3rd edition). Singapore: World Scientific.

Haidt, J. (2007). The New Synthesis in Moral Psychology. *Science, 316*(5827), 998-1002.

Haidt, J., Graham, J. (2007). When Morality Opposes Justice: Conservatives have Moral Intuitions that Liberals may not Recognize. *Social Justice Research, 20*(1), 1-19.

Haji, I. (2002). Compatibilist Views of Freedom and Responsibility. In R. Kane (Ed.), *The Oxford handbook of free will* (pp. 202-228). Oxford: Oxford University Press.

Hampshire, S. (1972). Spinoza and the Idea of Freedom. In S. P. Kashap (Ed.), *Studies in Spinoza, critical and interpretive essays* (pp. 310-331). Berkeley, CA: University of California Press.

Harari, Y. N. (2015). *Sapiens: a brief history of humankind* (First U.S. edition. ed.). New York: Harper.

Harris, S. (2012). *Free will* (1st Free Press trade pbk. ed.). New York: Free Press.

Hawking, S. (1996). *The illustrated A brief history of time* (Updated and expanded ed.). New York: Bantam Books.

Haynes, J.-D. (2011). Decoding and predicting intentions. *Annals of the New York Academy of Sciences*, 9-21.

Hegel, G. W. F. (1807/1977). Phenomenology of Spirit.

Hegel, G. W. F. (1816/1989). *Hegel's Science of Logic* (A. V. Miller, Trans.). Atlantic Highlands, NJ: Humanities Press International.

Hegel, G. W. F. (1821/1991). *Elements of the philosophy of right* (H. B. Nisbet, Trans. A. W. Wood Ed.). Cambridge; New York: Cambridge University Press.

Hegel, G. W. F. (1830/1970). *Hegel's Philosophy of nature; being part two of the Encyclopaedia of the philosophical sciences (1830)* (A. V. Miller, Trans.). Oxford: Clarendon Press.

Hegel, G. W. F. (1830/1971). *Hegel's Philosophy of mind: being part three of the 'Encyclopaedia of the philosophical sciences' (1830)* (A. V. Miller, Trans.). Oxford: Clarendon Press.

Heidegger, M. (1927/2008). *Being and time* (J. Macquarrie & E. S. Robinson, Trans.). New York: HarperPerennial/Modern Thought.

Heidegger, M. (1993). *Basic writings: from Being and time (1927) to The task of thinking (1964)* (D. F. Krell Ed. Rev. and expanded ed.). San Francisco, Calif.: HarperSanFrancisco.

Heidegger, M. (1995). *The fundamental concepts of metaphysics: world, finitude, solitude* (W. McNeill & N. Walker, Trans.). Bloomington: Indiana University Press.

Heidegger, M. (2010). *Die Grundbegriffe der Metaphysik: Welt - Endlichkeit - Einsamkeit*. Frankfurt am Mein: Klostermann.

Herculano-Houzel, S. (2009). The human brain in numbers: a linearly scaled-up primate brain. *Frontiers in human neuroscience, 3*, 31-31. doi: 10.3389/neuro.09.031.2009

Herculano-Houzel, S. (2012). The remarkable, yet not extraordinary, human brain as a scaled-up primate brain and its associated cost. *Proceedings of the National Academy of Sciences, 109*(Supplement 1), 10661. doi: 10.1073/pnas.1201895109

Herculano-Houzel, S., Avelino-de-Souza, K., Neves, K., Porfírio, J., Messeder, D., Mattos Feijó, L., . . . Manger, P. R. (2014). The elephant brain in numbers. *Frontiers in neuroanatomy, 8*, 46-46. doi: 10.3389/fnana.2014.00046

Herculano-Houzel, S., Ribeiro, P., Campos, L., Valotta da Silva, A., Torres, L. B., Catania, K. C., & Kaas, J. H. (2011). Updated neuronal scaling rules for the brains of Glires (rodents/lagomorphs). *Brain, behavior and evolution, 78*(4), 302-314. doi: 10.1159/000330825

Hobbes, T. (1650/1654/1812). *The treatise on human nature and that on liberty and necessity. With a suppl. to which is prefixed an account of his life and writings by the editor [P. Mallet]* (P. Mallet Ed.). London: J. Johnson and Co. St. Paul's Church Yard.

Hobbes, T. (1651/1994). *Leviathan: with selected variants from the Latin edition of 1668* (E. M. Curley Ed.). Indianapolis: Hackett Pub. Co.

Hoffmeyer, J. (2010). God and the World of Signs: Semiotics and the Emergence of Life. *Zygon, 45*(2), 367-390.

Hume, D. (1977). *An enquiry concerning human understanding: A letter from a gentleman to his friend in Edinburgh.* Indianapolis: Hackett Pub. Co.

Iacoboni, M. (2005). Understanding Others: Imitation, Language, Empathy *Mechanisms of Imitation and Imitation in Animals, vol. 1 of Perspectives on Imitation: From Mirror Neurons to Memes B2 - Mechanisms of Imitation and Imitation in Animals, vol. 1 of Perspectives on Imitation: From Mirror Neurons to Memes* (pp. 77-100). Cambridge, MA: MIT Press.

Illich, I. (1992). *In the mirror of the past: lectures and addresses, 1978-1990.* New York: M. Boyars.

Ismael, J. T. (2015). On Being Someone. In A. R. Mele (Ed.), *Surrounding free will: philosophy, psychology, neuroscience* (pp. 274-297). Oxford; New York: Oxford University Press.

James, W. (1884/2010). The Dilemma of Determinism *The Will to Believe: And Other Essays in Popular Philosophy* (pp. 168-207). [Auckland, N.Z.]: The Floating Press.

James, W. (1890/1981). *The principles of psychology* (Vol. 1). Cambridge, Mass.: Harvard University Press.

Jonas, H. (2001). *The phenomenon of life: toward a philosophical biology.* Evanston, Ill.: Northwestern University Press.

Kane, R. (2002). Introduction: The Countours of Contemporar Free Will Debates. In R. Kane (Ed.), *The Oxford handbook of free will* (pp. 3-41). Oxford: Oxford University Press.

Kane, R. (2007). Libertarianism *Four views on free will* (pp. 5-43). Malden, MA; Oxford: Blackwell Pub.

Kane, R. (2011a). Introduction: The Contours of Contemporary Free-Will Debates (Part 2). In R. Kane (Ed.), *The Oxford handbook of free will* (2nd ed., pp. 3-35). Oxford: Oxford University Press.

Kane, R. (2011b). Rethinking free will: new perspectives on an ancient problem. In R. Kane (Ed.), *The Oxford handbook of free will* (2nd ed., pp. 381-404). Oxford; New York: Oxford University Press.

Kant, I. (1787/1998). *Critique of pure reason* (P. Guyer & A. W. Wood, Trans.). Cambridge; New York: Cambridge University Press.

Kant, I. (1790/2000). *Critique of the power of judgment* (P. M. Guyer, Eric, Trans. P. Guyer Ed.). Cambridge: Cambridge University Press.

Kant, I. (1798/2006). *Anthropology from a pragmatic point of view* (R. B. Louden Ed.). Cambridge; New York: Cambridge University Press.

Kant, I. (1996). *Practical philosophy* (M. J. Gregor Ed.). Cambridge; New York: Cambridge University Press.

Kasting, J. F., & Siefert, J. L. (2002). Life and the Evolution of Earth's Atmosphere. *Science, 296*(5570), 1066-1068.

Kirk, G. S., Raven, J. E., & Schofield, M. (1983). *The presocratic philosophers: a critical history with a selection of texts* (2nd ed.). Cambridge; New York: Cambridge University Press.

Knausgaard, K. O. (2017). *Autumn* (I. Burkey, Trans.). New York: Penguin Press.

Knight, C., Dunbar, R., & Power, C. (1999). An evolutionary approach to human culture. In R. Dunbar, C. Knight & C. Power (Eds.), *The evolution of culture: an interdisciplinary view* (pp. 1-11). New Brunswick, N.J.: Rutgers University Press.

Kramer, S. N. (1963). *The Sumerians: their history, culture, and character*. Chicago: University of Chicago Press.

Kropotkin, P. (1921/2005). *Mutual Aid: A Factor of Evolution [Nook e-book]*. New York: Dossier Press.

Kuhn, T. S. (1962). *The structure of scientific revolutions*. Chicago: University of Chicago Press.

Kull, K. (1998). On semiosis, Umwelt, and semiosphere. *Semiotica, 120*(3/4), 299-310.

Kull, K., Stjernfelt, F., & Emmeche, C. (2002). A brief biosemiotic glossary. *Tartu Semiotics Library, 03*, 25-30.

Laidre, M. E., & Johnstone, R. A. (2013). Animal signals. *Current Biology, 23*(18), R829-R833. doi: https://doi.org/10.1016/j.cub.2013.07.070

Lamarck, J. B. P. A. d. M. d. (1809/1914). *Zoological philosophy: an exposition with regard to the natural history of animals* (H. S. R. Elliott, Trans.). London: Macmillan and Co., Limited.

Lee, N., Mikesell, L., Joaquin, A. D. L., Mates, A. W., & Schumann, J. H. (2009). *The interactional instinct: the evolution and acquisition of language*. Oxford ; New York: Oxford University Press.

Leibniz, G. W. (2005). *Discourse on Metaphysics and The Monadology* (G. R. Montgomery, Trans.): Dover Publications.

Libet, B. (2002). Do we have free will? In R. Kane (Ed.), *The Oxford Handbook of Free Will, Oxford University Press* (pp. 551-564). Oxford: Oxford University Press.

Libet, B., Gleason, C. A., Wright, E. W., & Pearl, D. K. (1983). Time of conscious intention to act in relation to onset of cerebral activity (readiness-potential): the unconscious initiation of a freely involuntary act. *Brain, 106*(3), 623-642. doi: 10.1093/brain/106.3.623

Liddell, H. S. (1947). The Experimental Neurosis. *Annual Review of Physiology, 9*(1), 569-580. doi: 10.1146/annurev.ph.09.030147.003033

Linden, E. (1974). *Apes, men, and language* (1st ed.). New York: Saturday Review Press.

Litt, T. (1961). *Mensch und Welt: Grundlinien einer Philosophie des Geistes*. Heidelberg: Quelle & Meyer.

Locke, J. (1689/1988). *Two treatises of government* (Student ed.). Cambridge; New York: Cambridge University Press.

Lorenz, K. (1977). *Behind the mirror: a search for a natural history of human knowledge* (1st American ed.). New York: Harcourt Brace Jovanovich.

Lotman, J. (1984/2005). On the semiosphere. *Sign Systems Studies, 33*(1), 205-229.
Lucretius. (1924). *On the Nature of Things* (W. H. D. Rouse, Trans. Vol. 181). ambridge, MA: Harvard University Press.
MacCallum, G. C., Jr. (1967). Negative and Positive Freedom. *The Philosophical Review, 76*(3), 312-334.
Magid, O. (2016). The Ontological Import of Heidegger's Analysis of Anxiety in 'Being and Time'. *Southern Journal of Philosophy, 54*(4), 440-462.
Maimonides, M. (1168/1948). *Eight Chapters (in Hebrew)* (S. Ibn Tivon, Trans.). Tel-Aviv: Rishonim Co.
Maimonides, M. (1191/2002). *Moreh nevukhim (The Guide to the Perplexed; in Hebrew)* (M. Schwarz, Trans.). Tel-Aviv: Universitat Tel-Aviv.
Manning, L. (2016). Future Mental Time Travel and the Me-self. In K. Michaelian, S. B. Klein & K. K. Szpunar (Eds.), *Seeing the future: theoretical perspectives on future-oriented mental time travel* (pp. 183-198): Oxford University Press.
Matthew, P. (1831). *On naval timber and arboriculture*.
McCabe, T. J. (1976). A Complexity Measure. *IEEE Transactions on Software Engineering, SE-2*(4), 308-320. doi: 10.1109/tse.1976.233837
McGrew, W. C. (2010). Chimpanzee Technology. *Science, 328*(5978), 579-580.
Mele, A. R. (2011). Free will and science. In R. Kane (Ed.), *The Oxford handbook of free will* (2nd ed., pp. 499-514). Oxford; New York: Oxford University Press.
Metcalfe, J. (2008). Evolution of metacognition. In J. Dunlosky & R. A. Bjork (Eds.), *Handbook of metamemory and memory* (pp. 29-46). New York, NY: Psychology Press.
Metzinger, T. (2003). *Being No One: The Self-Model Theory of Subjectivity*. Cambridge, Mass: A Bradford Book.
Metzinger, T. (2011). The No-Self Alternative. In S. Gallagher (Ed.), *The Oxford Handbook of the Self* (pp. 279-296). Oxford: OUP Oxford.
Michaelian, K., Klein, S. B., & Szpunar, K. K. (2016). The Past, the Present, and the Future of Future-Oriented Mental Time Travel. In K. Michaelian, S. B. Klein & K. K. Szpunar (Eds.), *Seeing the future: theoretical perspectives on future-oriented mental time travel* (pp. 1-18): Oxford University Press.
Midgley, M. (1995). *Beast and man: the roots of human nature* (Rev. ed.). London; New York: Routledge.
Midgley, M. (1996). *Utopias, dolphins, and computers: problems of philosophical plumbing*. London; New York: Routledge.
Milgram, S. (1963). Behavioral Study of obedience. *The Journal of Abnormal and Social Psychology, 67*(4), 371-378. doi: 10.1037/h0040525
Mill, J. S. (1859/2010). On LibertyOn Liberty and Other Essays (Amazon Kindle ed.): digireads.com.
Miller, S. L. (1953). A Production of Amino Acids under Possible Primitive Earth Conditions. *Science, 117*(3046), 528-529.
Moons, W. G., Mackie, D. M., & Garcia-Marques, T. (2009). The impact of repetition-induced familiarity on agreement with weak and strong arguments. *Journal of Personality and Social Psychology, 96*(1), 32-44. doi: 10.1037/a0013461
Nagasawa, M., Mitsui, S., En, S., Ohtani, N., Ohta, M., Sakuma, Y., . . . Kikusui, T. (2015). Oxytocin-gaze positive loop and the coevolution of human-dog bonds. *Science, 348*(6232), 333.

Nagel, T. (1974). What Is It Like to Be a Bat? *Philosophical Review, 83*(4 (Oct. 1974)), 435-450.

Nealon, J. T. (2015). *Plant Theory: Biopower and Vegetable Life*. Palo Alto: Stanford University Press.

New Genome Comparison Finds Chimps, Humans Very Similar at the DNA Level. (2005, 8/31/2005). Retrieved 12/27/2017, 2017, from https://www.genome.gov/15515096/2005-release-new-genome-comparison-finds-chimps-humans-very-similar-at-dna-level/

OpenStax. (2018). *Biology* Retrieved from https://cnx.org/contents/GFy_h8cu@11.2:fVAf83sY@13

Pennisi, E. (2016). Shaking up the Tree of Life. *Science, 354*(6314), 817.

Pepperberg, I. M. (2002). *The Alex studies: Cognitive and Communicative Abilities of Grey Parrots*. Cambridge, MA: Harvard University Press.

Pereboom, D. (2007). Hard Incompatibilism *Four views on free will* (pp. 85-125). Malden, MA ; Oxford: Blackwell Pub.

Pico della Mirandola, G. (1998). *On the dignity of man* (C. G. Wallis, P. J. W. Miller & D. Carmichael, Trans.). Indianapolis: Hackett Pub.

Pink, T. (2004). *Free will: a very short introduction*. Oxford; New York: Oxford University Press.

Plotinus. (1995). *Plotinus in seven volumes* (A. H. Armstrong, Trans.): Harvard University Press.

Popper, K. R. (2002). *The Logic of Scientific Discovery*: Routledge.

Price, T. D., Qvarnström, A., & Irwin, D. E. (2003). The role of phenotypic plasticity in driving genetic evolution. *Proceedings. Biological sciences, 270*(1523), 1433-1440. doi: 10.1098/rspb.2003.2372

Prothero, D. R. (2017). *Evolution: what the fossils say and why it matters* (Second edition. ed.). New York: Columbia University Press.

Putterman, T. L. (2006). Berlin's Two Concepts of Liberty A Reassessment and Revision. *Polity, 38*(3), 416-446.

Quammen, D. (2018). *The tangled tree: a radical new history of life* (First Simon & Schuster hardcover edition. ed.). New York, NY: Simon & Schuster.

Rödl, S. (2007). *Self-consciousness*. Cambridge, Mass.: Harvard University Press.

Rousseau, J.-J. (2002). *The social contract and The first and second discourses* (S. Dunn Ed.). New Haven: Yale University Press.

Russell, B. (1912/2004). *The problems of philosophy*. New York: Barnes & Noble Books.

Sartre, J.-P. (2007). *Existentialism is a humanism*. New Haven: Yale University Press.

Schneider, C. (2009). *Metaphysische Freiheit: Kohärenz und Theorie*. München: Philosophia.

Schopenhauer, A. (1839/1960). *Essay on the freedom of the will*. New York: Liberal Arts Press.

Schultze-Kraft, M., Birman, D., Rusconi, M., Allefeld, C., Görgen, K., Dähne, S., . . . Haynes, J.-D. (2016). The point of no return in vetoing self-initiated movements. *Proceedings of the National Academy of Sciences of the United States of America, 113*(4), 1080-1085. doi: 10.1073/pnas.1513569112

Scruton, R. (2017). *On human nature*. Princeton, NJ: Princeton University Press.

Sen, A. (1999). *Development as freedom* (1st. ed.). New York: Knopf.

Sen, A. (2002). *Rationality and freedom*. Cambridge, Mass.: Belknap Press.

Seyfarth, R. M., Cheney, D. L., & Marler, P. (1980). Monkey Responses to Three Different Alarm Calls: Evidence of Predator Classification and Semantic Communication. *Science, 210*(4471), 801-803.

Simard, S., Perry, D., Jones, M., D. Myrold, D., M. Durall, D., & Molina, R. (1997). Net transfer of C between ectomycorrhizal tree species in the field. *Nature, 388*, 579-582. doi: 10.1038/41557

Singer, P. (2009). *Animal liberation: the definitive classic of the animal movement* (Updated ed.). New York: Ecco Book/Harper Perennial.

Skinner, B. F. (1976). *Walden Two*. New York: Macmillan.

Smith, J. M., & Szathmary, E. (1997). *The Major Transitions in Evolution* Retrieved from http://ebookcentral.proquest.com/lib/ugalib/detail.action?docID=684606

Spinoza, B. d. (1677/1905). *Compendium grammatices linguae Hebraeae (in Hebrew)* (S. Rubin, Trans.). Kraków: Podgórze-Kraków, Drukiem i nakadem J. Plessnera i Ski.

Spinoza, B. d. (1677/1992). *The ethics; Treatise on the emendation of the intellect; Selected letters* (S. Shirley, Trans. S. Feldman Ed. 2nd ed.). Indianapolis: Hackett Pub. Co.

Spitzer, Nicholas C. (2015). Neurotransmitter Switching? No Surprise. *Neuron, 86*(5), 1131-1144. doi: https://doi.org/10.1016/j.neuron.2015.05.028

Stephenson, G. R. (1967). Cultural acquisition of a specific learned response among rhesus monkeys. In D. Starek, R. Schneider & H. J. Kuhn (Eds.), *Progress in Primatology* (pp. 279-288). Stuttgart: Fischer.

Strawson, P. (1982). Freedom and Resentment. In G. Watson (Ed.), *Free will* (pp. 59-80). Oxford; New York: Oxford University Press.

Stroop, J. R. (1935). Studies of interference in serial verbal reactions. *Journal of Experimental Psychology, 18*(6), 643-662.

Susanna, L. (2004). Heidegger's animal. *Phänomenologische Forschungen*, 57-81.

Szpunar, K. K., Spreng, N. R., & Schacter, D. L. (2016). Toward a Taxonomy of Future Thinking. In K. Michaelian, S. B. Klein & K. K. Szpunar (Eds.), *Seeing the future: theoretical perspectives on future-oriented mental time travel* (pp. 21-35): Oxford University Press.

Taylor, C. (1964). *The explanation of behaviour*. London, New York,: Routledge & Kegan Paul; Humanities Press.

Taylor, C. (2016). *The Language Animal: The Full Shape of the Guman Linguistic Capacity*. Cambridge, Massachusetts: The Belknap Press of Harvard University Press.

Taylor, J. S. (2005). Introduction. In J. S. Taylor (Ed.), *Personal autonomy: new essays on personal autonomy and its role in contemporary moral philosophy* (pp. 1-29). New York: Cambridge University Press.

Terrace, H. S., Petitto, L. A., Sanders, R. J., & Bever, T. G. (1979). Can an Ape Create a Sentence? *Science, 206* (23 November), 891-902.

Thompson, E. (2007). *Mind in life: biology, phenomenology, and the sciences of mind*. Cambridge, Mass.: Belknap Press of Harvard University Press.

Tibbetts, E. A., & Izzo, A. (2010). Social Punishment of Dishonest Signalers Caused by Mismatch between Signal and Behavior. *Current Biology, 20*(18), 1637-1640. doi: https://doi.org/10.1016/j.cub.2010.07.042

Timofeeva, O. (2018). *The History of Animals: A Philosophy*. London: Bloomsbury Publishing.

Trewavas, A. (2003). Aspects of Plant Intelligence. *Annals of Botany, 92*, 1-20.

Tversky, A., & Kahneman, D. (1974). Judgment under Uncertainty: Heuristics and Biases. *Science, 185*(4157), 1124-1131.

Uexküll, J. v. (1926). *Theoretical Biology* (D. L. MacKinton, Trans.). New York: Harcourt, Brace & Company, Inc.

Uexküll, J. v. W.-Y., Geoffrey. (2010). *A foray into the worlds of animals and humans: with A theory of meaning* (J. D. O'Neil, Trans. 1st University of Minnesota Press ed.). Minneapolis: University of Minnesota Press.

van Inwagen, P. (1982). The incompatibility of free will and determinism. In G. Watson (Ed.), *Free will* (pp. 47-58). Oxford; New York: Oxford University Press.

van Inwagen, P. (2000). Free Will Remains a Mystery. *Philosophical Perspectives, 14*, 1.

Vila, C., & Wayne, R. K. (1999). Hybridization between Wolves and Dogs. *Conservation Biology, 13*(1), 195-198.

Vygotsky, L. S. (1986). *Thought and language* (A. Kozulin, Trans. A. Kozulin Ed. Translation newly rev. and ed.). Cambridge, Mass.: MIT Press.

Walter, H. (2001). *Neurophilosophy of free will: from libertarian illusions to a concept of natural autonomy* (C. Klohr, Trans.). Cambridge, Mass.: MIT Press.

Ware, R. W., & Lopresti, V. (1975). Three-Dimensional Reconstruction from Serial Sections. In G. H. Bourne, J. F. Danielli & K. W. Jeon (Eds.), *International Review of Cytology* (Vol. 40, pp. 325-440): Academic Press.

Warren, K. J. (1995). The Power and the Promise of Ecological Feminism. In J. P. Sterba (Ed.), *Earth ethics: environmental ethics, animal rights, and practical applications* (pp. 231-241). Englewood Cliffs, N.J.: Prentice Hall.

Wegner, D. M., & Wheatley, T. (1999). Apparent mental causation: Sources of the experience of will. *American Psychologist, 54*(7), 480-492. doi: 10.1037/0003-066X.54.7.480

Weiner, J. (2005). Evolution in Action. *Natural History, 115*(9), 47-51.

Why Mouse Matters. (2017). Retrieved 12/26/2017, from https://www.genome.gov/10001345/importance-of-mouse-genome/

Wilkinson, A., & Huber, L. (2012). *Cold-Blooded Cognition: Reptilian Cognitive Abilities*.

Winfield, R. D. (2010). *Hegel and mind: rethinking philosophical psychology*. Basingstoke; New York: Palgrave Macmillan.

Winfield, R. D. (2011). *The living mind: from psyche to consciousness* (electronic edition (GoogleBooks) ed.). Lanham, Md.: Rowman & Littlefield Publishers.

Winfield, R. D. (2015). *The intelligent mind: on the genesis and constitution of discursive thought*.

Winfield, R. D. (2017). *Conceiving nature after Aristotle, Kant, and Hegel*. New York, NY: Springer Berlin Heidelberg.

Winfield, R. D. (2018). *Universal Biology after Aristotle, Kant, and Hegel: The Philosopher's Guide to Life in the Universe*: Springer International Publishing.

Wittgenstein, L. (2009). *Philosophical investigations* (G. E. M. Anscombe, P. M. S. Hacker & J. Schulte, Trans. Rev. 4th ed.). Chichester, West Sussex; Malden, MA: Wiley-Blackwell.

Woese, C. R., Kandler, O. M. L., & Wheelis, M. L. (1990). *Towards a Natural System of Organisms - Proposal for the Domains Archaea, Bacteria, and Eucarya* (Vol. 87).

Wohlleben, P. (2016). *The Hidden Life of Trees: What They Feel, How They Communicate—Discoveries from a Secret World* (T. Flannery, Trans.): Greystone Books.

Wolf, S. (2005). Freedom within Reason. In J. S. Taylor (Ed.), *Personal autonomy: new essays on personal autonomy and its role in contemporary moral philosophy* (pp. 258-274). New York: Cambridge University Press.

Wong, K. (2014). The 1 Percent. *Scientific American* (September), 100.

Woody, J. M. (1998). *Freedom's embrace*. University Park, Pa.: Pennsylvania State University Press.

Yudanin, M. (2013). Liberty and Freedom: The Relationship of Enablement *Applied Ethics: Risk, Justice, Liberty*: Center for Applied Ethics and Philosophy.

Yudanin, M. (2014). Can Positive Duties be Derived from Kant's Categorical Imperative? *Ethical Theory and Moral Practice*, 1-20. doi: 10.1007/s10677-014-9546-4

Zagzebski, L. (2002). Recent Work on Divine Foreknowledge and Free Will. In R. Kane (Ed.), *The Oxford Handbook of Free Will, Oxford University Press* (pp. 45-64). Oxford: Oxford University Press.

Zahavi, A. (1975). Mate Selection - A Selection for a Handicap. *Journal of Theoretical Biology, 53*(1), 205-214.

Zhang, H., & Wang, Y. (2011). Differences in hoarding behavior between captive and wild sympatric rodent species *Current Zoology, 5*(6), 725-730.

Иванов, В. I., Vyacheslav). (1999). *Избранные труды по семиотике и истории культуры (Selected papers on semiotics and the history of culture)* (Vol. 1). Moscow: Yazyki russkoy kultury.

Лобашев, M. L., Mikhail). (1961). Сигнальная наследственность (Signal inheritance). In M. Лобашев (Ed.), *Исследования по генетике (Research in genetics; in Russian)* (pp. 1-11). Leningrad.

Мамардашвили, M. M., Merab). (2011). *Сознание и цивилизация (Consciousness and Civilization; in Russian)*. Санкт Петербург: Азбука.

Чечулин, В. Л. С., V. L.). (2011). О различии этимологии слова "свобода" в русском и иных языках (On the difference between the etymology of the word "svoboda" (freedom) in Russian and other languages). *Приволжский научный вестник*(1), 44-50.

Notes

INTRODUCTION

1. This can pertain only to specific philosophical disciplines, not to philosophy as a whole: the latter examines its own foundations and thus establishes itself. In our example, the examination of the legitimacy of epistemology would be part of philosophical reasoning.

2. There are some exceptions to this rule, e.g., the work of Melvin Woody (Woody, 1998) and, to a degree, Daniel Dennett (see, for example, Dennett, 2003 and Dennett, 2015). However, the majority of the philosophical work on freedom still consists in the refinement of definitions and attacking alternative ways of defining what freedom is.

3. See *Critique of Practical Reason*, 5:48 in Kant, 1996

4. Chomsky, 2016, 24

5. René Descartes, 1642/1996, III:35

6. Discussed in more detail in Chapter 1, section *Accounts of freedom.*

7. Bookchin, 2005, 215

8. I used the review in Чечулин, 2011 for some etymological information, even though I find it hard to agree with author's analysis.

9. Also used to denote independence.

10. Kramer, 1963, 79-83 and specifically 79

11. Bookchin, 2005, 245

CHAPTER 1

1. This approach is a further development of the one proposed by Adler in *The Idea of Freedom* – see Adler & Institute for Philosophical Research (Chicago Ill.), 1958/1973, 20-27.

2. Adler & Institute for Philosophical Research (Chicago Ill.), 1958/1973, 21
3. Woody, 1998
4. Berlin, 2002, 32
5. Berlin, 2002, 32. Similar conception of freedom is advanced by J. S. Mill, even though he focuses on the societal aspect of freedom – see Mill, 1859/2010.
6. Berlin, 2002, 41, 169
7. See, for example, MacCallum, 1967; see also review of arguments for positive freedom in Putterman, 2006, 420-423.
8. Rousseau, 2002, esp. Book I, chapters 1, 4, 6, 7, and 8
9. *Critique of Practical Reason*, 5:33 – see Kant, 1996, 166
10. Berlin, 2002, 36n
11. Woody, 1998, 27
12. Hobbes, 1651/1994, 136 (Part II, Chapter XXI, Section 2)
13. Hegel, 1821/1991, §4 *Addition*.
14. Hegel, 1821/1991, §15, §15 *Addition*; contrary to the interpretation in Adler & Institute for Philosophical Research (Chicago Ill.), 1958/1973, 88-89
15. Mill, 1859/2010, Chapter 5
16. Schopenhauer, 1839/1960, 3
17. Schopenhauer, 1839/1960, 4-5, in the context of physical and intellectual freedom.
18. Wolf, 2005, 263
19. Harris, 2012, 17
20. Se, 1999, esp. xi-xii, 3, 10, and 17
21. Arendt, 1961, 151-154
22. See, for example, Kane, 2002, 5. This incorporates the definition of autonomy as personal self-governance (Beauchamp, 2005, 310).
23. See similar analysis in Woody, 1998, 29
24. Based on Schneider, 2009, esp. 7-8
25. See similar case in Woody's discussion of absolute freedom in Woody, 1998, 112
26. Schneider addresses these possibilities as *verwirklichbar*, realizable.
27. Schneider's *geistigen Individuum*; I believe that a translation more mundane that "spiritual" is in order here.
28. The "we" in "what we want."
29. Following Ayer, 1954/1982, 20-21
30. Dennett, 2015, 57-58
31. स्वतंत्र
32. Based on Cardona, 2014, 89-90
33. The discussion here is based on Dennett, 2015, 55-80 (Chapter 3 Control and Self-Control).
34. This interpretation is sometimes taken as obvious, not requiring any further discussion. See, for example, Ayer, 1954/1982, 15; Chisholm, 1982, 25; Frankfurt, 1969, 829. In both editions of the *Oxford Handbook of Free Will* which are nearly a decade apart the introduction notes this interpretation as well – see Kane, 2002 and Kane, 2011a.

35. I will address quantum indeterminacy later.

36. Wolf, 2005; Wolf analyses the case in terms of responsibility, yet, as the issue of responsibility is tied by her and others to the issue of freedom, it can be analyzed in terms of freedom as well.

37. I rely here on the presentation of Kane in Kane, 2002

38. Frankfurt, 1969

39. This specific example is mine.

40. Frankfurt, 1969, 836

41. More on that in Chapter 2, specifically in the section *Language*.

42. This is a major part of the argument of Harris, whose account popularizes the view that determinism is incompatible with freedom. See Harris, 2012, esp. 17-26.

43. A basis for this argument is Libet's research that discovered that by monitoring neural potential (Readiness Potential) that "wells up" before certain decision such decision can be predicted with high accuracy – see Libet, Gleason, Wright, & Pearl, 1983. For a more recent research with similar findings, see Haynes, 2011.

44. The review is based on Zagzebski, 2002

45. I altered the formulation of the argument in Zagzebski, 2002, 46-47, to avoid falling prey to the Alternative Possibilities problem and its solutions that have been discussed earlier.

46. Boethius, 525/1999, 119-128 (Book V, sections 3-4)

47. Maimonides, 1191/2002, 474-483 (volume 2, Chapter 17)

48. Maimonides, 1168/1948, 214 (Chapter 8)

49. Ayer, 1954/1982, 23

50. Ayer, 1954/1982, 22-23; see a similar argument in Midgley, 1995, xxxvii-xxxviii

51. Researches who continued Libet's program make a point to mention this. "[E]ven a complete, 100% prediction may not directly imply a causal link between the early predictive signals and the choice," notes Haynes (Haynes, 2011, 17). It was also found that human subjects have a sort of veto power over their decisions after readiness potential has been detected: when told that a potential for certain action had been detected in their brain functioning, subjects could change their mind and act differently up until ~200ms before the action (see Schultze-Kraft et al., 2016).

52. For a general exploration of this view as a commonly held position see Pink, 2004, esp. pages 22-25. It was considered as obvious by Kant, for example.

53. This is usually considered to require addressing options as falling under this or that concept, e.g., good or bad, proper or improper, etc. (see, for example, Davidson, 2004, Chapter 9 *What Thought* Requires, 137-138). For a more thorough treatment of rationality, see Chapter 2, section *Rational intelligence*).

54. Having preference toward pleasure vs. pain can and does lead to many choices in verbal and non-verbal animals and can be considered a common criterion for making choices. Singer, following Bentham, takes this as a starting point for animal ethics (Singer, 2009, 7 and elsewhere).

55. See a detailed exploration of non-rational choice in Chapter 2, section *Non-verbal animals: non-conceptual choice*.

56. Dennett, 1992/2014, 103

57. Even with Dennett's fictional yet useful self it is unclear how such self can be determining itself, similarly to a novelist writing a sequel – see Dennett, 1992/2014

58. Dennett, 2015, 88-89

59. Jonas in *Cybernetics and Purpose: A Critique* tell a similar story about a self-steering torpedo that shares our robot's relevant characteristics (Jonas, 2001, 115)

60. Artificial neural networks attempt to mimic the working of the central neural system in mammals by using computer programming. Neural network elements are programmatic models of neurons that process certain inputs according to a pre-defined formula and output a result. The program allows for weights of individual neurons' formulas to be modified based one whether the result or the processing of inputs matches the criterion set by a human operator utilizing the network. This way, a neural network can be "trained," or have a set of different weights assigned to its neurons, to provide a desired response to specific kinds of inputs, e.g., response 'cat' for pictures of cats rather than dogs, ants, or sunsets. This, by the network developing a weight configuration that will distinguish between pictures of cats and other pictures based on scanned color patterns. For example, when the network categorizes the pictures of cats correctly, the neurons that respond to the presence of whiskers and colors white, gray, orange, and black can gain greater weight than those that respond to the number of extremities and colors yellow, red, and green. A network "trained" this way will be able to process the next image and categorize it as a 'cat' or as 'not a cat.' (For intro to neural networks, see Graupe, 2013, esp. Chapter 1, 1-3, and Chapter 3, 1-16).

If one slides a rock down a wooden board and records the time it takes for it to get from top to bottom, one will discover that such time is negatively correlated with the number of time the rock slid over the board. This, because the board smoothens a bit each time the rock slides over it. Philosophically, the "training" of neural networks is much closer to the "training" of a wooden board than to the learning demonstrated by humans and other self-aware living entities. This is because in the latter cases the criterion of learning, its goals, and the ways to accomplish it are determined by the living organism, while with wooden boards and computers these are determined externally, by the programmer of the experimenter. Even in cases of simple classical conditioning, e.g., salivation of Pavlov's dogs at the hearing of a bell associated with feeding, the initial stimulus, satisfaction of hunger, was internally determined; same applies to the type of response, salivation.

In the last chapter I discuss the possibility for a certain constellation of neural networks to amount for more, yet these are waters yet to be charted.

61. First, of course, the world has to be conducive to the appearance and support of the choosing self, e.g., a living self, of being a biosphere. See Winfield, 2018, section 3: Nature as Biosphere, esp. 10-17 for details.

62. The notion of *Umwelt* has been suggested by Jakob von Uexküll in the early twentieth century and further developed in conjunction with Yuri Lotman's semiotics; see, for example, Kull, 1998 and Hoffmeyer, 2010. The concept keeps proving useful for the analysis of mind, as evident from recent work, e.g., Dennett, 2018, 77-84, 165-171 and elsewhere.

63. See similar discussion in Dennett, 2015, 80

64. A metaphor common to the Nyaya school of Indian philosophy – see Dasti, 2014, 112

65. An objection can be raised that some animals do change the world: hedgehogs and many birds build nests, ants and termites erect elaborate multi-story dwellings, beavers put together dams, etc. This, however, does not amount to intentionally changing the world. These activities are driven by instinctual urges, and the only choice ants, beavers, and other animal builders have is the exact locations and specific materials. Storks cannot decide not to build nests and pitch tents instead: this would require addressing the need for a dwelling regardless of particular sensory dimensions and then considering possible options to satisfy the criteria derived from such a need. Non-verbal animals are incapable of that.

66. For more discussion on freedom as enabler see Yudanin, 2013

67. Lorenz, 1977, 56

68. For the discussion on discursive metacognition, see section *Language and discursive metacognition* in Chapter 2

69. Human freedom is addressed in Chapter 2, section *Humans: rational freedom*: metacognition and the way language shapes the character of freedom is addressed in sub-section *Language*, the character of human intelligence – in sub-section *Rational intelligence*, and the resulting nature of human freedom – in sub-section *The nature of human freedom*. Animal and human exercise of choice will be considered in section *Animal choice vs human freedom: a comparison*.

70. Woody, 1998, 14

71. Leibniz, 2005, 52 (*Monadology*, §32)

72. See similar argument described in Dennett, 2015, 91

73. Hawking, 1996, 106

74. See similar argument in Campbell, 2011, 23-25, where he also relies on van Inwagen and Strawson

75. Following Balaguer, 2010, 9

76. Harris, 2012, 16-17

77. 'Laws of physics' here refer to the observed regularities of the material world that are independent of agent's volitions. See discussion in van Inwagen, 1982, 48-49 for more details.

78. Following van Inwagen, 1982, 47

79. For a review of different definition of determinism as pertaining to the question of free will, see Walter, 2001, 15-16

80. Fischer, Kane, Pereboom, & Vargas, 2007, 2

81. There are philosophers who equate this ability with 'could have acted otherwise' (see, for example, van Inwagen, 1982, 49-50). It ignores the considerations specified in section *Why not other criteria*; however, incompatibilism would treat the ability to choose between alternatives that is up to the agent in the same way as it addresses the power to act otherwise.

82. Based on Kane, 2007, 10-11

83. Van Inwagen, 1982, 54

84. Kane, 2007, 6

85. Kane, 2007, 5; Aristotle's account of freedom in the *Nicomachean Ethics* also emphasizes the aspect of 'up to us' (ἐφ' ἡμῖν) for cases of free decision. See, for example, Book III, 1112a31(Aristotle, 1984).

86. Kane, 2007, 14-15; see also Kane, 2011b

87. Kane, 2007, 23

88. Kane, 2007, 24

89. Pink, in his popular textbook account of freedom, emphasizes the commonsensical aspect of libertarianism, including its incompatibilist corollary (Pink, 2004, 12-14).

90. It is not clear that we should take quantum indeterminacy into consideration in this case, as its impact on macro-level events that are subject to assumed human choice seems questionable.

91. See Kane, 2007, 9

92. Yet not necessarily equal probability, of course.

93. See Pink, 2004, 83-89 (exercise problem); Kane, 2007, 9-13; Kane, 2011a, 19-24

94. Kane, 2007, 25

95. See *Critique of Practical Reason*, 5:48 in Kant, 1996

96. Based on Kane, 2007, esp. 26-33. Similar account appears in Kane, 2011b.

97. Kane, 2007, 26

98. Kane, 2007, 26

99. Kane appeals to the notion of neural networks and the increased effect of neural events that the conditions of chaos bring, yet these are secondary to the argument.

100. Kane, 2007, 29-30

101. Kane, 2007, 33

102. Kane, 2007, esp. 35-36

103. Van Inwagen, 2000

104. Balaguer, 2010

105. Balaguer, 2010, 97

106. See, for example, Kane, 2007, 29

107. Lucretius, 1924, 115 (lines 251-262)

108. For a more detailed critique of Kane's position along the same lines, see Pereboom, 2007, esp. 101-110

109. Spinoza, 1677/1992, 50 (I, Pr. 28)

110. For our purposes, Spinoza's conception of God can be seen as identical to the conception of Nature as an all-encompassing lawful deterministic whole, of which everything is a part. Further, we can narrow it to the material.

111. Interpreting Spinoza's argument as deterministic seems not entirely correct, as he has a causality of reason paralleling physical causality (for a similar interpretation see Hampshire, 1972). However, here I am using Spinoza's argument for deterministic causality in order to explain the foundations of determinism.

112. Bédécarrats, Chen, Pearce, Cai, & Glanzman, 2018. The researchers transplanted RNA from trained sea snails into untrained sea snails and observed the latter demonstrating conditioned behavior learned by the former.

113. Libet et al., 1983

114. Haynes, 2011

115. Mele, 2011, 508. For a popular account of free will as epiphenomenal see Harris, 2012, esp. 8-11. For a more detailed account that suggests seeing free will as epiphenomenal based on experimental research, see Wegner & Wheatley, 1999, as well as review in Mele, 2011, esp. 508-514.

116. The way our nervous system is organized is genetically determined: its elements are encoded in the DNA.

117. This is a bit different from the definition of delusion accepted in clinical psychology, which is states that delusion is a "false belief based on incorrect inference about external reality that is firmly held despite what almost everyone else believes and despite what constitutes incontrovertible and obvious proof or evidence to the contrary" (DSM-5, "Glossary of Technical Terms," 2013). I omit the appeal to the commonality of belief in defining delusions, since I believe that it is not essential to characterizing this phenomenon: after all, beliefs that are contrary to the clear counterevidence can be quite common.

118. Libet, 2002

119. Schultze-Kraft et al., 2016

120. Woody makes a similar point in Woody, 1998, 203-204

121. Dennett offers similar analysis using the sorites fallacy – see Dennett, 2015, 92. See also Walter, 2001, 66.

122. Metzinger, 2003, 5

123. Using the terms as in Albahari, 2006, 51

124. Verse 279. *The Dhammapada*, 2010, 55

125. This interpretation of the Buddhist doctrine of no-self (anatta / अनत्त) is presented in Chapter 3 of Albahari, 2006, 50-80

126. In Metzinger, 2011

127. Metzinger, 2011, 283

128. Metzinger, 2011, 284

129. Metzinger, 2011, 285

130. Metzinger, 2011, 285, 285n2. I would note that Rödl presented a coherent philosophical theory of first-person perspective that is compatible with embodied self in his *Self-Consciousness* (Rödl, 2007).

131. Metzinger, 2011, 285

132. Metzinger, 2003, 1 and elsewhere

133. See Metzinger's treatment of intuitiveness in Metzinger, 2011, 287-289

134. See, for example, Kane, 2002; Haji, 2002; Fischer, 2007; Kane, 2011a, and more.

135. Hobbes, 1650/1654/1812, 116

136. Hobbes, 1650/1654/1812, 125

137. Hobbes, 1650/1654/1812, 155; see also *Leviathan*, Chapter XXI, [1] (Hobbes, 1651/1994, 136)

138. Hobbes, 1651/1994, 136

139. Hobbes, 1650/1654/1812, 119

140. Hobbes, 1651/1994, 136

141. Hobbes, 1651/1994, 120

142. Hobbes, 1651/1994, 122

143. Hobbes, 1651/1994, 123

144. Hobbes, 1651/1994, 124

145. Hobbes, 1650/1654/1812, 133-134

146. Hobbes, 1650/1654/1812, 116

147. See, for example, Fischer, 2007, 51

148. Hobbes, 1650/1654/1812, 141

149. The will for action here is different from Libet's readiness potential in that it is conscious.

150. The interpretation of Spinoza's interpretation of free will here follows the argument in Hampshire, 1972. I am indebted to professor Edward Halper for discussing this interpretation in his seminar.

151. Part II, Axiom 2 (Spinoza, 1677/1992, 64)

152. II, Ax. 3 (Spinoza, 1677/1992, 64); II, Definition 3

153. II, Proposition 7 (Spinoza, 1677/1992, 66)

154. II, Pr. 7, Proof (Spinoza, 1677/1992, 64)

155. II, Pr. 7, Scholium (Spinoza, 1677/1992, 67)

156. II, Def. 4 (Spinoza, 1677/1992, 63)

157. IV, Pr. 6, Sch. (Spinoza, 1677/1992, 191-192)

158. On reason causing strong emotions, see V, Pr. 6 and Pr. 6 Sch. (Spinoza, 1677/1992, 205-206)

159. IV, Pr. 73, Sch. (Spinoza, 1677/1992)

160. *Bhagavad Gita: The Song Selestial*, 1965, Part III

161. See, for example, brief review in Haji, 2002, 202-203

162. Skinner describes how it can be done in *Walden Two* (Skinner, 1976). This approach makes the existing moral discourse obsolete, yet it does assign importance to praise and blame on different, deterministic grounds.

163. Hume, 1977, 63

164. Strawson, 1982, 63; see also analysis in Haji, 2002, 203-210

165. See, for example, review in Haji, 2002, 210-217

166. See, *Critique of Practical Reason*, 5:96-97 in Kant, 1996

167. *Critique of Practical Reason*, 5:97 in Kant, 1996

168. James, 1884/2010

169. Van Inwagen, 2000, 9-10

170. Dennett, 2015, 54

CHAPTER 2

1. Schneider, 2009, 57

2. James, 1890/1981, p279; the italics are James's.

3. In this sense, it is similar to Hegel's essential identity – see Hegel, 1816/1989, 411

4. See similar conception of the self in Thompson, 2007, esp. 48-49 and 487n7

5. Kant, 1790/2000, 5:371; Akademie edition references are used here and elsewhere.

6. Somewhat wider than Aristotle's material causality – see Physics II.3 and Metaphysics V.2, Aristotle, 1984

7. Metaphysics V.2, Aristotle, 1984

8. Kant, 1790/2000, 5:371

9. Heidegger, 1995, 232

10. Kant, 1790/2000, 5:374
11. Kant, 1790/2000, 5:371
12. Although Kant treats the products of environmental material's processing as educts, substances separated from a mixture in which it already exists, both his description and our current knowledge suggest chemical changes rather than non-chemical separation.
13. Mostly human, yet we should not ignore the production, as opposed to mere use, of simple tools by other animals – see, for example, McGrew, 2010 on tool use in great apes, and Godfrey-Smith, 2016, 64 on tool use in octopuses.
14. Jonas, 2001, 75-76
15. Jonas, 2001, 79
16. This distinction does not preclude artificially created life from being a life, whatever physical form it takes, yet it does distinguish between programmable entities and living beings.
17. Kant, 1790/2000, 5:371-372
18. Kant, 1790/2000, 5:373
19. Kant, 1790/2000, 5:372
20. Kant, 1790/2000, 5:373; see also Jonas, 2001, 79-80
21. Jonas, 2001, 5:372; see also Jonas, 2001, 79
22. Jonas, 2001, 79
23. Kant, 1790/2000, 5:375
24. Kant, 1790/2000, 5:371
25. Jonas, 2001, 44
26. Dennett, 1995, 75-76 and elsewhere.
27. Both referred to as cold-bloodedness.
28. Jonas, 2001, 75
29. Thompson, 2007, 44-45 and elsewhere.
30. Thompson, 2007, 43
31. Thompson, 2007, 106
32. Jonas, 2001, 45. A more detailed account of the evolution of life as relevant to the question of freedom is provided in Chapter 4.
33. Popper, 2002, esp. 17-20
34. This might be seen as somewhat similar to Thomas Kuhn's ideas on the development of science – see Kuhn, 1962
35. More on that in Chapter 4.
36. "Soul must be a substance in the sense of the form of a natural body having life potentially within it" (Aristotle, 1986, 412a21-21)
37. Aristotle, 1986, 415b10-12
38. Aristotle, 1986, 415b15-16
39. Aristotle, 1986, 415a23-415b1
40. See sub-section *Life as self* above.
41. Aristotle, 1986, 424b22-426a1, 427a18, 432a15-29 and elsewhere in book III of *De Anima*.
42. Aristotle, 1986, 429a10-430b26
43. This is evident from Aristotle, discussion of human's ability to distinguish between "a magnitude and what it is to be a magnitude, and between water and what

it is to be water, and so in many other cases" (Aristotle, 1986, 429a10-12) – between a thing and its form.

44. Wittgenstein, 2009, §§19, 23
45. Wittgenstein, 2009, §241
46. See Wittgenstein, 2009, §20 for an example.
47. Wittgenstein, 2009, *Philosophy of Psychology – A Fragment (former Part II)*, §1
48. Wittgenstein, 2009, *Philosophy of Psychology – A Fragment (former Part II)*, §1
49. Agamben, 2015, especially part III
50. Agamben takes the term from Plotinus (Plotinus, 1995, I.4, 3). See similar treatment in Illich, 1992, 227.
51. See Wittgenstein, 2009, §§241-243
52. Agamben, 2015, 54; italics are author's.
53. Agamben, 2015, 62
54. For a contemporary textbook account of operational conditioning, see Alcock, 2001, 100-101
55. More on that in Chapter 5
56. Specifically, the third section of the *Philosophy of Nature* (Hegel, 1830/1970) and the *Philosophy of Mind* (Hegel, 1830/1971), as well as Winfield's interpretation and further development of Hegel's approach in *The Living Mind* (Winfield, 2011) and *The Intelligent Mind* (Winfield, 2015)
57. For the purposes of analyzing forms of life as forms of freedom, all what can be said about plants applies to simpler types of life, namely fungi, bacteria, and archaea. I will focus on plants because they are evolutionarily closer to the next step – non-verbal animals.
58. This was noted already by Aristotle in *On Plants*: see Aristotle, 1984, 816b1, 10-15. See also Jonas, 2001, 99
59. See Aristotle, 1984, 816b1, 6-7, and Hegel, 1830/1970, §345 and §345R(emark).
60. Wohlleben, 2016, 7
61. See, for example, Simard et al., 1997
62. Wohlleben, 2016
63. See, for example, Brenner et al., 2006
64. Alpi et al., 2007, 136
65. Hegel, 1830/1970, §337, §345Z(usatz).
66. Hegel, 1830/1970, §344, §351Z
67. Hegel, 1830/1970, §344Z, §347Z
68. Hegel, 1830/1970, §346a
69. Hegel, 1830/1970, §348Z
70. Hegel, 1830/1970, §§346a, 347
71. Hegel, 1830/1970, §348Z
72. Agamben, 2015, 29-30
73. Spinoza, 1677/1905, Chapter 20, 79-80
74. Aristotle, 1984, 815b10, 20-30; see also Jonas, 2001, 99, 102
75. Aristotle, 1984, 815b, 10, 23

76. Hegel, 1830/1970, §337
77. Aristotle, 1984, 815b10, 2-4
78. Jonas, 2001, 99, 102
79. Sponges who lack nervous system can hardly be seen as motile; jellyfish lack brain yet have some semblance of centralized control provided by a nerve ring.
80. Jonas, 2001, 101
81. Jonas, 2001, 102
82. Jonas, 2001, 103
83. Hegel, 1830/1970, §337
84. Hegel, 1830/1970, §350Z
85. Hegel, 1830/1970, §355Z
86. Hegel, 1830/1970, §359
87. Hegel, 1830/1970, §360
88. Hegel, 1830/1970, §361
89. Hegel, 1830/1970, §368
90. Aristotle, 1984, *De Anima* III 4, 429a10, 12
91. Hegel, 1830/1971, §389
92. Winfield, 2010, 9
93. See similar argument in Winfield, 2010, 8-9
94. Winfield, 2015, 181
95. Kant, 1787/1998
96. Aristotle, 1984 , *De Anima* III 4, 429a10, 24
97. Aristotle, 1984, *De Anima* II 1, 412^{a}27-28, 412^{b}4-5
98. For tropism in plants, see discussion in Hegel, 1830/1970, §§337, 334, 334 Zusatz, §345 Zusatz. For a comparative discussion of plant tropism and animal subjectivity, see Winfield, 2011, 78-82.
99. See Winfield, 2011, 9
100. This does not necessarily entail the ability to conceptualize subjectivity and represent it to oneself, which would be a characteristic of discursive mind.
101. Winfield, 2011, 81
102. The distinction between signaling systems and language will be drawn later in this chapter.
103. Winfield, 2011, 81
104. Winfield, 2015, 26-28
105. Winfield, 2011, 81; Winfield, 2015, 28
106. Winfield, 2015, 28
107. Re-purposing existing neurophysiological resources seems to be an evolutionary mechanism that can play at various evolutionary junctions. See, for example, section *Metacognition and language* in Part II, Chapter 5.
108. Winfield, 2015, 16
109. Winfield, 2015, 20
110. Winfield, 2015, 10-11
111. Winfield, 2015, 38-39
112. Winfield, 2015, 54
113. Szpunar, Spreng, & Schacter, 2016, 22

114. C. Taylor, 1964, 9; see also 27-29

115. On empathy in animals, specifically primates, see, for example, de Waal, Macedo, Ober, & Wright, 2006. On empathic behaviors in rats see Church, 1959 and Bartal, Decety, & Mason, 2011.

116. See, for example, Iacoboni, 2005

117. De Waal et al., 2006, 25-26

118. De Waal et al., 2006, 33-36

119. De Waal et al., 2006, 31-32; see also 37 for review of other ToM research.

120. Flombaum & Santos, 2005

121. Bugnyar & Heinrich, 2006

122. Winfield, 2011, 201-204; Winfield, 2015, 8, 108

123. See more in-depth analysis in Winfield, 2011, 201-211 and specifically section 9.5

124. For a textbook account, see Alcock, 2001, esp. 299-315

125. See Terrace, Petitto, Sanders, & Bever, 1979

126. Clayton calls such communication systems "referential languages" as opposed to human languages (Clayton, 2009)

127. Terrace et al., 1979, 895

128. James, 1890/1981, 109

129. Lorenz, 1977, 56

130. For the discussion of instincts in humans, see section *Urges, instincts, and emotions* in Part II, chapter 5. The continuum of freedom, where the degree of freedom is inversely proportional to organism's dependence on environmental inputs, has been addressed by Clayton in his Gradualist Theory of Freedom as the emergence of freedom (Clayton, 2009, esp. 81-98).

131. James, 1890/1981, 22-23

132. Wilkinson & Huber, 2012

133. Hegel, 1816/1989, 117. Language will be discussed in section *Language* below; normativity will be discussed in sections *Language* and *Rational intelligence*.

134. For a similar analysis see C. Taylor, 1964, 17

135. Stephenson, 1967. This experiment apparently gave rise to an urban legend about an experiment with monkeys and cold shower, where each subsequent generation was taught to avoid going up the ladder to retrieve a banana because the first generation of the participants was exposed to this unpleasant circumstance as a result of scaling the ladder.

136. Brosnan & de Waal, 2003

137. Liddell, 1947

138. Winfield, 2015, 57-62, esp. 59

139. Dennett, 2018, 84

140. Agamben after Heidegger in Agamben, 2015, 89-91

141. Agamben, 2015, 90

142. This will be addressed in the next chapter.

143. The genetic endowment of all mammals is quite similar. For example, humans and rats share some 85% of protein-coding regions ("Why Mouse Matters," 2017). With chimpanzees the number is 96-99%, depending on whether we take into consideration DNA insertions and deletions ("New Genome Comparison Finds Chimps, Humans Very Similar at the DNA Level," 2005).

144. Fitch, de Boer, Mathur, & Ghazanfar, 2016; Dunn, 2018
145. "Bigger Brains: Complex Brains for a Complex World."
146. Terrace et al., 1979
147. Linden, 1974, 9-29
148. Linden, 1974, 77-135
149. In a sense, any mental activity is reflexive, as it is aware of itself (Winfield, private communication). The term is used here to describe conscious reflexivity.
150. Nagasawa et al., 2015
151. Aboutness has been analyzed in relation to truth following Russell's suggestion that truth always relates to something else, to cognitions that are outside true or false beliefs (see Russell, 1912/2004, 89-97 and esp. 90). Truth, however, is merely one case of the aboutness of language: all utterances are about something, and language is not limited in terms of what it is about. See further discussion in this section for details.
152. It is important to note that the question "is it good to love?" is of the same order as the question "is it valid to ask questions about love?" – both address mental contents that do not result directly from processing sensory data but from considering other metal contents.
153. For similar treatment of thinking, see Hegel, 1830/1971, §§465-468 and Winfield, 2015, 181-203 and esp. 181-185
154. Chomsky, 2016, 4
155. Chomsky, 2007, esp. 15-16; Chomsky, 2016, 16-17
156. Symbols will be discussed later.
157. For a brief review, see Laidre & Johnstone, 2013
158. Chomsky, 2016, 4. This is not a much-berated private language that develops without any input from the environment – such language has been discredited by Wittgenstein a while ago (Wittgenstein, 2009, §§ 256-283). The I-language develops based on sensory inputs, images obtained from the environment, and is expressed using grammar learned from others. Since language is used not only for making sense of the world by an individual but also as means of communication (as opposed to animal signaling – communication of emerged meanings), the meanings are mostly public – they are anchored by public use (see Winfield, 2015, 108), and those that are not yet public have a potential to become such.
159. Chomsky, 2016, 4
160. See, for example, Avargues-Weber & Giurfa, 2015. The authors claim that their results demonstrate conceptual learning in bees, yet the most that can be inferred is that bees might develop, under certain circumstances, simple cognitive maps of their environment.
161. This is, in fact, what the philosophy of mind and the current work are trying to do.
162. Winfield, 2015, 77-78; see also studies in animal conditioning and chimpanzee use of human sign language cited throughout this work, e.g., Linden, 1974
163. Winfield, 2015, 77; see also Hegel, 1830/1971, §455, §455Z
164. Hegel, 1830/1971, §456
165. Winfield, 2015, 80-81
166. Addessi, Mancini, Crescimbene, Padoa-Schioppa, & Visalberghi, 2008

167. Winfield, 2015

168. See Winfield, 2015, 75. Vygotsky, reviewing Köhler's studies, also notes that while chimps enjoy playing with colors, there is no indication that there is any representative intent in their creations (Vygotsky, 1986, 72-73). More recently, de Waal, while reviewing some impressive attempts to consider ape art seriously and even sell it successfully in the arts market, concludes that "[t]he main dividing line between ape and human art is representation. In spite of isolated claims of apes' producing recognizable images, human art seems to be unique in its ability to depict the reality that surrounds us" (de Waal, 1999).

169. This and other expression have been observed and documented in a variety of settings, including by Darwin in *The Expression of Emotion in Men and Animals* (C. Darwin, 1872/2009), in a variety of animals, including in social insects (see, for example, Tibbetts & Izzo, 2010)

170. For the externalization of aggressive intent in canines and felines, see, for example, C. Darwin, 1872/2009, 116-145)

171. Winfield, 2015, 84; Hegel, 1830/1971, §458

172. Vygotsky, 1986, 223

173. Hegel, 1830/1971, §460; Winfield, 2015, 92-93

174. Vygotsky, 1986, 223

175. Knausgaard, 2017, 215; see his essay *Silence*, 215-217

176. Knausgaard, 2017, 215

177. The role of verbal memory will be presented here only briefly; for in-depth discussion, see Hegel, 1830/1971, §§461-464 and Winfield, 2015, 99-107

178. Winfield, 2015, 102-103

179. As opposed to the famous Kantian dictum; Winfield, 2015, 106

180. Winfield, 2015, 106; Hegel, 1830/1971, §464

181. Hegel, 1830/1971

182. This section is based on Winfield, 2015, 108-118

183. Wittgenstein, 2009, §§19, 23

184. Some animals, most notably rodents, hoard food for later consumption. Yet this does not constitute planning for the future; rather, such behaviors are instinctual. Some aspects of the hoarding behavior are learned, yet its core constitutes an inherited behavioral pattern. See Zhang & Wang, 2011 for discussion.

185. For Bischof-Köhler hypothesis, see, for example, Bischof, 1980, 57-58

186. Michaelian, Klein, & Szpunar, 2016, 10

187. Winfield, 2015, 16

188. Winfield, 2015, 84

189. See, for example, Atance, 2008

190. Szpunar et al., 2016, 24

191. From λείπεται δὴ πρακτική τις τοῦ λόγον ἔχοντος, active (practical) life of the element that has *logos* Aristotle, 1984, *Nicomachean Ethics,* I:7, 1098a3-5

192. C. Taylor, 2016, 338

193. Sen, 2002, 4

194. Sen, 2002, 4, 19-26

195. Manning, 2016, 184, 186

196. This is discussed in Chapter 2, section *Non-verbal animals: non-conceptual choice*, subsection *Mind.*

197. Manning, 2016, 193

198. Winfield, 2015, 106; see also Hegel, 1830/1971, §464

199. Winfield, 2015, 10

200. Winfield, 2015, 13; see also 118

201. Winfield, 2015, 123

202. Garfield, 2014, esp. 173-174 and 179-180

203. Garfield, 2014, 174

204. Winfield, 2015, 124

205. See, for example, accounts of Berns's fMRI scanning of dogs' brains in setting of food reward and praise in Berns, 2017.

206. Pico della Mirandola, 1998, 4-5

207. In the preface to his *Anthropology from a Pragmatic Point of View*, Kant echoes similar view by defining anthropology as the study of a human being not only of "what *nature* makes of the human being "but also "the investigation of what *he* as a free-acting being makes of himself, or can and should make of himself" (Kant, 1798/2006, 3). Here the treatment of the subject itself, namely the normative dimension, demonstrates the acting of conceptual freedom.

208. The impact of human freedom on evolution will be discussed in Part II, Chapter 5, section *The emergence and evolution of discursively intelligent animals.*

209. Winfield, 2011, 58

210. Arendt, 1968, 75. Arendt is using a quotation from Droysen's *Historic* (1882), surmising that the Greek quote he uses sounds Aristotelean.

211. See discussion on instinct earlier in this chapter in section *Instincts and choice.*

212. Scruton, 2017, 11; see also 12-13

213. Eagleton, 2000, 95

214. Clayton, 2009, 77

215. Stroop, 1935

216. Sartre, 2007, 29

217. This fallacy is a further development of the original Availability heuristic suggested by Tversky and Kahneman (Tversky & Kahneman, 1974)

218. Moons, Mackie, & Garcia-Marques, 2009

219. Illich, 1992, 222

220. This seems to be the process Descartes following in his *Mediations*, where he postulates existence of the immaterial soul based on conceiving it and the body in a certain way: René Descartes, 1642/1996, esp. *First* and *Second Meditations.*

221. Lorenz, 1977, 230

222. De Waal et al., 2006, 33-36

223. the sense of deliberation used her is similar to Aristotle's deliberative choice – see *Nicomachean Ethics*, Book III, 2 (Aristotle, 1984, 1111b1-1112a15)

224. More on that see in Yudanin, 2014

225. Milgram, 1963

226. See Church, 1959. For more recent studies in rat empathy see Bartal et al., 2011. I am indebted to Kirill Voytsel who brought these studies to my attention.

227. It is questionable whether any human society can persist while not only rejecting human freedom at the level of theory but following through on such belief in practice, yet this discussion goes beyond the scope of the current work.

228. For the role of society in limiting and enabling freedom see Yudanin, 2013

229. Dennett, 1995, 75-76 and elsewhere

230. De Waal et al., 2006, 31-32; see also 37 for review of other Theories of Mind research

231. I am indebted to Kirill Voytsel for his enlightening thoughts on the nature and importance of hope.

232. For a review of philosophical approaches to hope see Bloeser & Stahl, 2017

233. See, for example, Kant's analysis of hope in the *First Critique*: Kant, 1787/1998, A805-806/B833-834.

234. I am not familiar with research on animal response to radically unfamiliar stimuli; this area might be waiting to be explored.

235. *Bhagavad Gita: The Song Selestial*, 1965, Chapter 11, 65-71

PART II

1. See Chapter 2, section *Representing consciousness – the birthplace of choice.*
2. See Chapter 2, section *Humans: rational freedom.*

CHAPTER 3

1. Winfield, 2018, 2
2. Dobzhansky, 1964, 451
3. See Chapter 2, section *Representing consciousness – the birthplace of choice.*
4. Animal signaling will be discussed in this Chapter 5, section *The emergence and evolution of non-verbal animals.*
5. Culture is not a mere collection of behavioral patterns, including communications by signaling, but, on any definition, involves customs, ways of doing things that carry a normative aspect (Winfield, 2015, 122). This requires language, as it has been argued for in Chapter 2, section *Language and discursive metacognition*. In a wider sense, culture can be seen as a system of signs that can be interpreted by the members of this culture's *semiosphere*, the space within which the signs that comprise it can be generated and interpreted (see Lotman, 1984/2005, esp. 208). Interpretation of signs, as opposed to symbols, is not possible without linguistic intelligence which non-verbal animals lack. Culture will be discussed in more detail in section *The human universe: myths and culture.*
6. For a thorough exploration of the concept of design space, see Dennett, 1995, 85-145

7. The relationship between natural selection and freedom will be explored in more detail in Chapter 5

8. Animals frequently have been observed responding to challenges to which they could not have been conditioned or equipped with appropriate instinctual behavioral patterns – see, for example, Köhler's apes, rats that find their way through mazes, dogs that run to a doggy door after seeing a squirrel through a window in another room, etc., and similar examples. Animal learning is discussed throughout Chapter 5, section *The emergence and evolution of non-verbal animals.*

9. Later this would be couched in terms of the development of language in Chapter 5, section *The emergence and evolution of discursively intelligent animals.*

10. Addressed in detail in Chapter 2, section *Humans: rational freedom.*

11. See Chapter 2, section *The Basic Property and Merge: recursive and reflexive.*

12. C. Darwin, 1859/2017, xxvi

13. See section *The emergence and evolution of non-verbal animals* later in Chapter 5

14. See section *Hegelian evolution* in Chapter 4. Bookchin also points to directedness of the evolutionary process, even though his scope is much wider than the one I am addressing here – see Bookchin, 2005, 460-461, 466, 468

CHAPTER 4

1. See, for example, Dobzhansky, 1955, 1

2. The link between freedom and life is established in Chapter 2.

3. In what follows, I summarize the definitions suggested in various sources: Erasmus Darwin's *Zoonomia* (E. Darwin, 1794, esp. section XXXIX *Of Generation*), Dobzhansky (Dobzhansky, 1956), and the modern textbook account in OpenStax, 2018, Unit 4: *Evolutionary Processes.* At this point I will try to avoid committing the understanding of evolution to any specific theory.

4. This is how evolution is frequently defined, as change of species over time that gives rise to new species through gradual change and related genetic distancing (see, for example, OpenStax, 2018, 18.1 Understanding Evolution); speciation, however, seems to be a consequence of evolution, not something that is identical with it. Reproductional isolation in this context means that the members of one group, or species, do not commonly mate with the members of another. This can be due to genetic distance that is large enough to prevent mating or the lack of desire to attempt to mate with another group; it can also simply be a consequence of a geographic isolation, e.g., when a natural barrier exists between the members of an animal or plant group. The old definition of distinct species as not producing fertile offspring was recently found outdated, as some related groups of animals that had been considered for long different species were found to interbreed and produce fertile offspring, e.g., all species of the genus *Canis* (see, for example, Vila & Wayne, 1999); many other species produce fertile offspring as well, an offspring that can mate with the non-mixed mother species and thus contribute new DNA inherited from another species to it (see Pennisi, 2016).

5. Dobzhansky, 1964, 449

6. The argument here is rooted in Jonas's (Jonas, 2001, 38-58) and Winfield's (Winfield, 2018, 66-75) analysis of Darwinism. A somewhat different angle will be explored later in section *Hegelian evolution.*

7. See Chapter 2, section *Life as self.*

8. This is achieved both through mutations and later – the reshuffling of genetic material in sexual reproduction.

9. Kirk, Raven, & Schofield, 1983б 303-304, DK 375, 380

10. See Physics, Book II.8 (Aristotle, 1984, 198b17-33)

11. See E. Darwin, 1794, esp. section XXXIX *Of Generation.*

12. See review in Kirk et al., 1983, 95-97

13. In his *Zoological Philosophy* (Lamarck, 1809/1914)

14. In his *On the Origin of Species* (C. Darwin, 1859/2017)

15. The three principles had been briefly described, in a somewhat unstructured manner, in the summary of Chapter IV of the *Origin* (C. Darwin, 1859/2017, 112-113), and since then have been given many compact formulations. See, for example, OpenStax, 2018, 18.1 Understanding Evolution. For the discussion of different formulations and their merits see Godfrey-Smith, 2007.

16. See section *Natural selection.*

17. These segments of the DNA, inherited from the evolutionary ancestors, are usually referred to as "junk DNA," and they amount to 80-97% of the genetic code (Prothero, 2017, 104-107). Some researchers are unsure whether junk DNA is not playing any role, or we are yet to discover the role it plays (see, for example, modern textbook account in OpenStax, 2018, 17.2 Mapping Genomes).

18. Same amino acids, organism's building blocks, frequently can be produced by different segments of the DNA (OpenStax, 2018, 15.1 The Genetic Code).

19. OpenStax, 2018, 11.1 The Process of Meiosis, 13.1 Chromosomal Theory and Genetic Linkage.

20. For a classical statement of the Modern Synthesis see Dobzhansky, 1955, esp. chapters 4 and 5. See also critical review from the vantage point of 2017 in Prothero, 2017, 100-106.

21. For discussion and examples see Prothero, 2017, 107-111

22. See reviews in Weiner, 2005, and Prothero, 2017, 121-126

23. See OpenStax, 2018, 16.3 Eukaryotic Epigenetic Gene Regulation for details.

24. See section *Natural selection* later.

25. The subject of choice's impact on evolution and the evolution of the ability to choose will be discussed in Chapter 5

26. This will be discussed in Chapter 5, specifically in section *The emergence and evolution of discursively intelligent animals.*

27. Thompson, 2007, 176, 202

28. For a brief discussion on dominant and recessive genes, see OpenStax, 2018, 2.3 Laws of Inheritance. For a modern understanding of dominance and recessivity, see Chapter 13. Modern Understanding of Inheritance.

29. Lamarck observed that the use of a certain organ will enlarge and strengthen it, while the disuse make it shrink and lose function – this can be seen in multiple cases,

from prehensile vs. unused tails to pelvic bones in elephants vs. whales. He suggested that the environmental changes lead to the change in the way animals operate, thus causing to use some organs more than others. By passing the resulting modifications to their offspring, animals improve its chances to survive. This, of course, only if both mates in sexually reproducing animals possess similar modifications – otherwise they cancel each other. It is this environmental influence that, for Lamarck, makes for the diversity in animal species that otherwise would be much more uniform in their evolutionary march from lesser to greater complexity. See Lamarck, 1809/1914, Chapter VII, esp. 107-108, 113, 124.

30. Prothero, 2017, 103
31. Thompson, 2007, 191
32. Prothero, 2017, 104
33. OpenStax, 2018, 14.6 DNA Repair.
34. See previous section, *Variation*, and review in Thompson, 2007, 176-177
35. Thompson, 2007, 177
36. For this type of intergenerational transmission of behavioral characteristics Lobashev coined the term "signal inheritance" (Лобашев, 1961). However, since it is not utilizing the usual biological inheritance mechanisms, I will refer to it as social learning or social transmission.
37. There isn't much research in this area. See, however, Stephenson, 1967.
38. For more on procedural knowledge, see Fantl, 2017
39. See Winfield, 2015, 108-109, 115. For a detailed discussion of language see Chapter 2, section *Language and discursive metacognition.*
40. Similar ideas have been expressed before Darwin, and he duly acknowledges this in *An Historical Sketch* right after the introduction to the *Origin* (C. Darwin, 1859/2017, xxi-xxxi)
41. Matthew, 1831, 365
42. Dobzhansky, 1956, 62
43. This is typical for discursively intelligent organisms, as I will argue later.
44. See Kropotkin, 1921/2005, esp. chapters I and II
45. For an alternative and popular explanation that has been deemed the Handicap principle see Zahavi, 1975
46. Dobzhansky, 1956, 65
47. Dobzhansky, 1956, 65
48. Thompson, 2007, 204
49. See section *Evolving biological entities* in Chapter 5
50. See Chapter 5
51. Lamarck, 1809/1914, 130; the argument is developed on 129-131
52. For a brief discussion on the origin of life see section *Starting down the evolutionary road* in Chapter 5
53. See, for example, Winfield, 2018, 81-82. Dennett addresses this subject as well in Dennett, 1995, 155-156
54. Winfield, 2018, 80; see also 80-81
55. In Aristotelean terms, i.e., in terms of the progression from nutritive to animal to rational soul.

56. Winfield, 2018, 81

57. This definition of complexity is essentially quantitative and resembles the one suggested by McCabe (McCabe, 1976); for a brief discussion on the genetic definition of complexity see Smith & Szathmary, 1997, 5-6. However, it does not reduce evolutionary development to simple addition of parts and functions, as they and their interaction bring about more abilities; see discussion in Jonas, 2001, 24.

58. These options will be discussed in the next section, *Hegelian evolution.*

59. See section *The emergence and evolution of non-verbal animals* in Chapter 5

60. For a more detailed discussion on how language and metacognition work together, see section *Metacognition and language* in Chapter 5

61. See section *The emergence and evolution of discursively intelligent animals* in Chapter 5

62. Hegel, 1830/1970, §249 Remark.

63. Hegel, 1830/1970, §249 Remark.

64. Jonas, 2001, 38-63

65. Winfield, 2018, esp. 57-87

66. Hegel, 1830/1970, §249 Remark.

67. Hegel, 1816/1989, 128. See Timofeeva, 2018, 97 for similar analysis of the contradiction in animal nature and evolutionary development.

68. For the definition of complexity, see section *Increase in complexity.*

69. In what follows, I develop the philosophical aspect of what has been briefly described in the previous section, subsection *Increase in complexity.*

70. Winfield, 2018, 67; see also Jonas, 2001, 44

71. See review of evidence in Prothero, 2017, 94-127. In some individuals the genes that have been inherited from the evolutionary ancestors yet kept inactive in their species got activated, thus leading to multi-toed horses or humans with a tail (*Ibid.,* 104-107).

72. See Agamben, 2015, 54 and elsewhere.

73. Heidegger, 1993, 228-231; the name is built on the analogy of the Greek *ekstasis* (ἔκστασις), with the ἐκ to emphasize the out-standing nature of human existence in the world.

74. Heidegger, 1993, 234

75. Heidegger, 1993, 262; see similar interpretation of Heidegger in Timofeeva, 2018, 137-138

76. The necessary conditions for self-determined choice in terms of animal capacities are specified in Chapter 2, section *Representing consciousness – the birthplace of choice.*

CHAPTER 5

1. Miller, 1953; the experiment tested the hypothesis suggested by Oparin. It should be noted that creating the conditions for the emergence of life is not equivalent to the creation of life: life's essence is in creating itself through metabolism, it

cannot be created externally. Still, life first emerges from inorganic material, not as an artifact, but as a contingent concatenation of macromolecules within a permeable membrane within which autopoiesis can proceed.

2. See, for example, Kasting & Siefert, 2002, 1066

3. Winfield, 2018, 57

4. Winfield, 2018, 10, 12

5. See Chapter 2, section *Life as self*

6. Also, complex structures like the organic macromolecules of which life of Earth consists tend to break over time. Therefore, the second law of thermodynamics requires them to be open systems in order to not give in to the pressures of entropy (Dennett, 1995, 127; on the entropy at the chemical level see Chapter 6 *Transformation of Chaos* in Atkins, 1984, 107-125).

7. Lamarck suggested, following the ancients, that life in its simplest form emerges directly out of non-living matter with the help of heat and moisture (see Chapter VI in Lamarck, 1809/1914, 236-248, esp. 237). Hegel wrote about spontaneous generation as *generatio aequivica*, generation that is not of the being of the same kind (Hegel, 1830/1970, §341 Zusatz).

8. Winfield, 2017, 387

9. Hegel, 1830/1970, §335 Zusatz.

10. Winfield, 2017, 389-392

11. See Thompson, 2007, 44-45 and elsewhere.

12. Miller, 1953

13. See, for example, Cairns-Smith, 2009. I am following here an interpretation suggested in Eisen & Konchok, 2018, 77-78.

14. Eisen & Konchok, 2018, 78

15. Winfield, 2017, 393

16. Winfield, 2017, 396

17. See discussion in Chapter 2, section *Life as self*, as well as Thompson, 2007, 91, 97-107 and Winfield, 2018, 29-31

18. See Chapter 2, section *Life as providing a potential for the development of self-control and self-awareness*.

19. See discussion on slime mold and its significance later, in section *From unicellular to multicellular organisms*.

20. See, for example, Jonas's *Is God a Mathematician? The Meaning of Metabolism* in Jonas, 2001, 64-98

21. See, for example, Winfield, 2018, 42-43

22. See section *Hegelian evolution* in Chapter 4

23. Winfield, 2018, 78

24. Winfield, 2018, 130

25. This development will be discussed in the section *The emergence and evolution of discursively intelligent animals*.

26. I am ignoring here the implications of the theory of relativity and, specifically, the notion of the speed of light as the highest speed of the spread of any impact, as this is not important for the discussion.

27. Uexküll, 1926, 70

28. Uexküll, 1926, 77, 126, 128; Uexküll, 2010, 42. Von Uexküll's indicator, or carrier of significance, is referred to by Heidegger as "disinhibitor" (*das Enthemmende*) – perhaps, a better way to emphasize the nature of the indicator as triggering organism's activity; see Agamben, 2004, 51 for discussion.
29. See Nagel, 1974
30. Uexküll, 2010, 44-46. This example will prove instrumental in analyzing human freedom as well.
31. Uexküll, 2010, 50, 53
32. Uexküll, 2010, 43
33. Heidegger, 1995, 169-366
34. Heidegger, 1995, 284; see also 287, 292, 304, etc.
35. See summary in Heidegger, 1995, 284
36. Мамардашвили, 2011, 11
37. Мамардашвили, 2011, 15
38. Heidegger, 1995, 186-211
39. Uexküll, 2010, 219; citing Theodor Litt in Winthrop-Young's *Afterword*; for more on Litt's treatment of human vs. animal Umwelts, see Litt, 1961, esp. 28-30
40. Uexküll, 2010, 220; citing Theodor Litt in Winthrop-Young's *Afterword.*
41. Uexküll, 2010, 142
42. Uexküll, 2010, 158
43. Uexküll, 2010, 140
44. For the discussion on life as semiosis see Kull, 1998
45. Kull, Stjernfelt, & Emmeche, 2002, 29
46. For the treatment of signs in this book see section *From associative imagination to human language* in Chapter 2. For the definition of a sign in biosemiotics see Kull et al., 2002, 30.
47. Hoffmeyer, 2010, 373
48. Hoffmeyer, 2010, 386
49. For a brief discussion of *noosphere*, see de Chardin, 1955/2008, 180-184
50. Kull, 1998; the term originates with Lotman (Lotman, 1984/2005)
51. See section *Increase in complexity* in Chapter 4
52. Hoffmeyer, 2010, 377; Hoffmeyer also calls it "semiotic freedom" (Ibid.).
53. Hoffmeyer, 2010, 378
54. Hoffmeyer, 2010, 379
55. Agamben, 2004, 40-41
56. See section *The emergence and evolution of discursively intelligent animals.*
57. Winfield, 2018, 45; see also Jonas, 2001, 86
58. Heidegger, 1927/2008, 228-235/H186-H191
59. Magid, 2016, 411
60. For Heidegger this would not be a serious argument, yet it seems apt in the current context.
61. Heidegger, 1927/2008, 232-236/H187-H192
62. Heidegger, 1927/2008, 232/H188
63. Heidegger, 1927/2008, 232/H188

64. Based on the explanation of plasticity in Price, Qvarnström, & Irwin, 2003, 1433

65. Prothero, 2017, 108. This type of plasticity is called *heterochrony*, as it consists in changing the timing of biological transformations based on environmental triggers, and is an example of *neotony*, retention of organism's juvenile body form while successfully reproducing.

66. See, for example, Simard et al., 1997

67. See review in Costandi, 2016, esp. 33-114 (chapters 3-7)

68. For a brief definition of procedural and declarative knowledge, see Chapter 4, section *Heredity*.

69. Dennett, 2018, 164

70. Costandi, 2016, 101-114

71. This process, of course, is not bullet-proof, can have and does have unintended consequences, and is prone to create as much evil as good due to people's lack of knowledge or ideological positions regarding the environment. However, it is this process that is responsible to the unique viability of the humankind and our proliferation to all parts of the world.

72. Winfield, 2018, 73-74

73. Dennett, 1995, 85

74. Quammen, 2018, 8

75. C. Darwin, 1859/2017, 102-103

76. See review in Quammen, 2018, 3-34, 188

77. Quammen, 2018, 41-44, 52-53

78. Quammen, 2018, 52-54, 61

79. Woese, Kandler, & Wheelis, 1990

80. For Dennett's design space, see Dennett, 1995, 124-145

81. Dennett, 1995, 106

82. See discussion on the transformation of pneumococcus in Quammen, 2018, 215-218

83. See discussion on genetic crossover earlier in this chapter, section *Biological evolution*, subsection *The principles of evolution - variation.*

84. The up-to-date textbook account of HGT can be found in OpenStax, 2018, 20.3 Perspectives on the Phylogenetic Tree.

85. For the review of relevant research, e.g., that of Lynn Margulis, see Quammen, 2018, 113, 115, 119, 127, 131, 133-135, 142, 149-250, and 349. Microbiologists see the origin of eukaryotes in the incorporation of bacteria by archaea (Cox, Foster, Hirt, Harris, & Embley, 2008; OpenStax, 2018, 22.2 Structure of Prokaryotes).

86. Quammen, 2018, 249

87. See review in Quammen, 2018, 227-229, 231, 239, 325-326, 340-342, 349, 351-352. Some 8% of human DNA is said to originate in retroviruses.

88. Doolittle, 1999, 2142

89. Quammen, 2018, 252-253

90. See, for example, Goldenfeld & Woese, 2007; for discussion, see Quammen, 2018, 252-254

91. Quammen, 2018, 290-291

92. The related evolutionary developments will be discussed in the section *The emergence and evolution of discursively intelligent animals.*

93. This will be discussed in detail in the section *The emergence and evolution of discursively intelligent animals.*

94. This, of course, should not be read as a declaration of optimism, as human history clearly demonstrates.

95. For the simpler and more ancient unicellulars, e.g., bacteria like *Mycoplasma mycoides*, this is the only way of individuation. With evolution, the unicellulars become more complex. They develop multiple internal structures, or organelles, intercellular communication through molecular messaging, and even some rudimentary movement, as it can be seen in *Paramecia*. All these contribute to their individuation, yet the main characteristic of a unicellular organism is still its biological unity: it has no distinct character besides that, not even a stable genetic code due to the prolific horizontal gene transfer.

96. Smith & Szathmary, 1997, 203

97. Slime mold is an informal umbrella name for several eukaryotic organisms that can live either as independent unicellular entities or form multicellular structures. The observations addressed here come from OpenStax, 2018, 23. Introduction, 23.2 Characteristics of Protists, and 23.3 Groups of Protists, as well as from Eisen & Konchok, 2018, 30, 32, 46-48, 50. See also Quammen, 2018, 383.

98. OpenStax, 2018, 23.2 Characteristics of Protists.

99. OpenStax, 2018, Chapter 23. Introduction.

100. Agamben, 2004, 38; see also 15

101. In terms of the life form type, fungi and plants can be characterized together.

102. Dennett, 2015, 41-42

103. See section *Variably restricted evolutionary paths* for the discussion on the design path.

104. More on that see later in the section that discusses animal development: *The emergence and evolution of non-verbal animals.*

105. In terms of their relation to the evolution of freedom, plants and fungi can be classified together. Thus, in what follows I would refer to them as plants or a plant life form.

106. Kasting & Siefert, 2002, 1066; Quammen, 2018, 144-145, 293-294

107. Carnivorous plants like the Venus fly trap are no exception: they get their nutrition from what comes into the immediate contact with them rather than chasing prey or otherwise moving to get food, thus maintaining the same principle of immediate access to nutrition.

108. Hegel, 1830/1970, §346a; Winfield, 2018, 106

109. Winfield, 2018, 93-94; see also 100

110. The use of the term here is particularly apt, as rhizome refers to the subterranean plant stem that sends out roots horizontally. I am using it in the meaning suggested by Nealon in his interpretation of Deleuze and Guattari (Nealon, 2015, Chapter 4, esp. 84-86).

111. I am following here the argument similar to that of Nealon (Nealon, 2015, 90)

112. See, for example, Nealon, 2015, esp. 12 and 30, for review and references.

113. Trewavas, 2003, esp. 6; see also 10

114. For the critique of the argument of plant unity from biological positions, see Firn, 2004, 345-346

115. Hegel, 1830/1970, §344 Zusatz; Winfield, 2018, 112. Carnivorous plants also metabolize immediately rather than chasing their prey.

116. Trewavas, 2003, 3; see also 1-4.

117. Winfield, 2018, 94

118. For the discussion on the different types of environments see the section *Starting down the evolutionary road*, subsection *World* earlier in this chapter.

119. Winfield, 2018, 134

120. Trewavas, 2003, 7

121. Trewavas, 2003, 7

122. For the critique of Trewavas's account of plant plasticity as memory and learning, see Firn, 2004, esp. 348

123. This is why vegetative life is discussed today as a criterion of meaningful human death, where its physiological correlates, the lack of brain activity, figure as the ultimate indication. See Agamben, 2004, 15; cit. from page 38.

124. Ismael, 2015

125. Ismael, 2015, 278

126. Jonas, 2001, 131; Winfield, 2018, 130-131

127. See Hegel, 1830/1970, §§350-376, and Winfield, 2018, 121. See also Jonas's *To Move and To Feel: On Animal Metabolism* (Jonas, 2001, 99-107); Lamarck too sees motility, sentience/irritability, and self-control as essential to the animal form of life (Lamarck, 1809/1914, 53).

128. Here referring to the simple urges, yet later developing into more complex feelings.

129. Hegel, 1830/1970, §337

130. Hegel, 1830/1970, §350 Zusatz, §350

131. Winfield, 2018, 107

132. Winfield, 2018, 121

133. For a brief discussion on hopeful monsters, see Prothero, 2017, 107-108

134. This section greatly benefited from the graduate seminar in the philosophy of mind that was taught by Edward Halper, University of Georgia, in the fall of 2013.

135. See, for example, Aristotle's *De Anima*, Book II, Chapter 3 (Aristotle, 1984, 414a29-415a14). Hegel adds to that the practicality of feeling, or directionality of emotion in the *Philosophy of Mind* (Hegel, 1830/1971, §§471-472 and esp.473); while Hegel's discusses emotion in the context of humans, his argument can apply to non-verbal animal as well – see discussion in Winfield, 2018, 131.

136. Aristotle, 1984, 414b2-6

137. For discussion on pain in invertebrates, see Elwood, 2011. For a more recent discussion on pain in animals that does not arrive at a definite conclusion yet provides interesting examples, see Adamo, 2016, esp. 76.

138. Heidegger, 1995, 242

139. Midgley, 1995, 20, 24

140. Midgley, 1995, 52

141. Midgley, 1995, 53; see also 52-55

142. Heidegger, 1995, 264 and elsewhere in Part Two of *The Fundamental Concepts of Metaphysics.*

143. Midgley, 1995, 53

144. Dennett, 2015, 11-12 and elsewhere: Dennett seems to be quite fond of this insect.

145. Alcock, 2001, 118.

146. See a somewhat similar argument in Winfield, 2018, 156

147. See Chapter 2, section *Instincts and choice.*

148. See Chapter 2, section *Representing consciousness – the birthplace of choice.*

149. Dennett, 2018, 164

150. Hegel, 1830/1970, §351, §351 Zusatz; Winfield, 2018, 142-143

151. Different lists of emotions have been suggested, e.g., by Aristotle in the *Nicomachean Ethics*, Book II, section 5 (Aristotle, 1984, 1105b21-22) and by Descartes in Part II of *The Passions of the Soul* (René Descartes, 1989, 50-101)

152. Aristotle, 1984, 1105b23-24

153. Arguably, there is no feeling pain without also being able to feel pleasure and vice versa.

154. Winfield, 2018, 139; this is different from ethical love that involves recognition of rights and duties of self and other and thus requires discursive intelligence.

155. This will be discussed later, in section *The emergence and evolution of discursively intelligent animals.*

156. In this Chapter 3, section *Philosophical understanding of the evolution of freedom: the suggested course.*

157. For a report on experiments with dogs' preference toward owners vs. food and an interesting discussion, see Chapter 7 in Berns, 2017 (137-158)

158. In many cases the discovery of the way instincts require the comparison of stored mental content to the intuited one involves substituting the object of the instinctual pattern by something else. For example, Konrad Lorenz inserted himself into the newly hatched goslings' view in order to have them *imprint* on him and then follow him as if he were their mother. As a result, these goslings would waddle after the scientist everywhere while growing up. See de Waal, 2016, 37-40. For a discussion on imprinting and other examples of this phenomenon, see Lorenz, 1977, 78-80.

159. Dobzhansky, 1956, 52

160. Costandi, 2016, 1, 147

161. Costandi, 2016, 2

162. Costandi, 2016, 73-78

163. Costandi, 2016, 37, 42-44, 54-61, 67, 71-83, 148

164. Costandi, 2016, 13

165. Costandi, 2016, 13, 152; see review in Spitzer, 2015. Most experiments in this area have been conducted on rats.

166. Costandi, 2016, 102-104; Walter, 2001, 117, 225

167. Walter, 2001, 115-122

168. Barth, 2018, 60

169. Barth, 2018

170. Costandi, 2016, 154

171. Later, I will argue that similar processes can characterize social and cultural functioning.

172. See discussion in de Waal, 2016, 51-53. For earlier discussion on transmitting habits among animals, see Dobzhansky, 1955, 340-341.

173. De Waal, 2016, 154-156

174. De Waal, 2016, 320

175. Human culture will be discussed in more detail in subsection *The human universe: myths and culture*.

176. See Chapter 2, section *Representing consciousness – the birthplace of choice*.

177. Ware & Lopresti, 1975

178. Bode et al., 1973

179. Dupeyroux, Viollet, & Serres, 2019

180. Herculano-Houzel et al., 2011

181. Herculano-Houzel, 2012

182. Herculano-Houzel, 2009

183. Herculano-Houzel et al., 2014

184. Godfrey-Smith, 2016, 32. For a brief survey of the Ediacaran fauna, see Prothero, 2017, 173-181.

185. Prothero, 2017, 1 (chart)

186. See, for example, a remarkable summary of a multi-year study of Grey African parrots in Pepperberg, 2002

187. Godfrey-Smith, 2016, 41

188. Godfrey-Smith, 2016, 52-58, 64, 73, 99-100, 109, 191

189. For a discussion on divergent and convergent evolution, see OpenStax, 2018, 18.1 Understanding Evolution.

190. Timofeeva, 2018, 86-87

191. The examples here are taken from de Waal, 2016, 237-240

192. Heidegger, 1995. "das Nichthaben *im* Habenkönnen is gerade das *Entbehren*, die *Armut*." (Heidegger, 2010, §50 / 309)

193. Heidegger, 1995, 247; see also 197-198 and 248

194. Susanna, 2004, 56

195. Metacognition is addressed in Chapter 2, section *Humans: rational freedom*, sub-section *Language*.

196. Heidegger, 1995, 269, 274; see also 284. I addressed this thesis earlier.

197. Heidegger, 1995, 238, 249, 253, 255, 257, 267

198. Agamben, 2004, 68

199. Heidegger, 1995, 241-246

200. A chimp named Vicki, when required to sort photographs into categories of humans and animals, did so correctly. When confronted with her own picture, she placed it without hesitation in the human pile, on top of Eleanor Roosevelt. The picture of her father, however, went with horses and elephants (Linden, 1974, 49-50).

201. Linden, 1974, 8

202. Midgley, 1995, 18

203. *Offen* yet not *offenbar* (Agamben, 2004, 55). While the English rendering of *offenbar* as *disconcealed* is not the most elegant, I follow here the translation that has become standard in the literature.

204. Timofeeva, 2018, 145

205. Heidegger, 1995, 247-248

206. OpenStax, 2018, 9.1 Signaling Molecules and Cellular Receptors.

207. OpenStax, 2018, 9.1 Signaling Molecules and Cellular Receptors, 9.3 Response to the Signal.

208. OpenStax, 2018, 45.7 Behavioral Biology: Proximate and Ultimate Causes of Behavior.

209. Laidre & Johnstone, 2013, 831

210. OpenStax, 2018, 45.7 Behavioral Biology: Proximate and Ultimate Causes of Behavior.

211. For the investigation of how representing consciousness enables choice see Chapter 2, section *Non-verbal animals: non-conceptual choice*, sub-section *Representing consciousness – the birthplace of choice*.

212. Laidre & Johnstone, 2013, 830; Darwin addressed the subject in relative detail in *The expression of the emotions in man and animals* (C. Darwin, 1872/2009)

213. For the discussion on symbols, see Chapter 2, section *From associative imagination to human language*.

214. For the initial research see Seyfarth, Cheney, & Marler, 1980; see also de Waal, 2016, 107-108

215. See Pepperberg, 2002

216. Linden, 1974, esp. 3-188

217. Godfrey-Smith, 2016, 127. See also 109-133

218. Wong, 2014, summarizing research

219. De Waal, 2016, 221; pretty much the whole book is the description of research that points in this direction with one notable exception – human language (106-107)

220. De Waal, 2016, 106

221. See Lee, Mikesell, Joaquin, Mates, & Schumann, 2009, 4-8 for an overview and the rest of the book – for research and argumentation.

222. Lee et al., 2009, 4-5, 151-166

223. Lee et al., 2009, 4, 11-54

224. Lee et al., 2009, 3

225. De Waal, 2016, 107

226. Metcalfe, 2008, 29, 30, 34-35

227. Metcalfe, 2008

228. Metcalfe, 2008, 30

229. I am echoing here Kant's famous dictum about the blindness of the categories without intuitions, while acknowledging the Hegelian insight that reason can work with the contents of its own making.

230. These studies are not new – see, for example, Bontempi, Demir, Destrade, & Jaffard, 1999. For a more recent review, see Costandi, 2016, 54-61.

231. Costandi, 2016, 22-27

232. Heidegger, 1995, 309

233. Aristotle, 1984, *De Anima*, 430a27-29

234. Heidegger, 1995, 314

235. Heidegger, 1995, 337

236. This development is addressed in detail in Chapter 2, section *From associative imagination to human language*. Here I consider its evolutionary aspect.

237. Harari, 2015, 4

238. I am not addressing the question of whether there are genuinely right or wrong ways to treat each other – I think there are. I am just pointing out that the ways we do treat each other are of our own making.

239. As argued earlier, humans have heritable general tendencies and inclinations yet not instincts. For a discussion on instincts, see section *Urges, instincts, and emotions* earlier in this chapter.

240. Uexküll, 2010, 220, *Afterword* by Winthrop-Young.

241. Berns, 2017, 137-158

242. Heidegger, 1995, 193

243. Uexküll, 2010, 158

244. Nagel, 1974, 411

245. See a wonderfully passionate and intellectually interesting treatment of this subject in Coetzee, 2016, 34-35

246. See Chapter 2, esp. section *Humans: rational freedom*, for a discussion on human freedom and its differences from animal choice.

247. Heidegger, 1995, 277

248. Uexküll, 2010, 220, *Afterword* by Winthrop-Young.

249. Midgley, 1996, 11

250. Иванов, 1999, 349, 530

251. See, for example, de Waal, 2016, 320

252. Boyd & Richerson, 2005, 6

253. See, for example, Distin, 2011, 11

254. This paragraph presents a view of information that is somewhat different from the common one. For comparison, see Floridi, 2010 and Distin, 2011, 11-24.

255. Stephenson, 1967; discussed in more detail in Chapter 2, section *Animal culture and morality?*

256. See section C. Self-alienated spirit (*B: Der sich entfremdete Geist; die Bildung*) in the *Phenomenology of Spirit* (Hegel, 1807/1977, §§484-595, esp. §§484-493)

257. See analysis in Bookchin, 2005, chapters 3 and 4

258. See, for example, Warren, 1995, esp. 232-234

259. Agamben, 2004, 26

260. *>als<-Struktur*. It is also the condition of the possibility of truth and falsity, and thus of reason. See Heidegger, 1995, 314-315.

261. This analysis is rooted in Hegel, 1807/1977, 487

262. Agamben, 2004, 29

263. "Der Geist macht sich selbst zu dem, was er ist." Litt, 1961, 27, presenting Hegel's position.

264. See, for example, Boyd & Richerson, 2005, 8; Knight, Dunbar, & Power, 1999, 2; and, for a more refined but principally similar view, Dennett, 2018, esp. chapters 5 (76-101) and 9-11 (176-247)

265. For an example, see Boyd & Richerson, 2005, 19-34

266. See beehive moral scales in J. Haidt, 2007 and J. Haidt, Graham, J., 2007

267. Litt, 1961, 26

268. Мамардашвили, 2011, 8

269. Heidegger, 1927/2008, 230/H186, 232-233/H187-H188. This would not apply to non-human animals as they would not be able to evaluate the way of being in the world: relation to the way of being in the world, e.g., to anxiety, is non-sensory.

270. Heidegger, 1927/2008, 232/H188

271. Costandi, 2016, 155

272. This has a direct implication on the utility of the IQ tests that are still in vogue: they measure capacities that are more important to primates and pre-technological humans than to us today.

273. Bookchin, 2005. The whole book attempts to trace the roots of humanity's choice of hierarchy and suggest an alternative.

274. Heidegger, 1995, 312; see also 309

275. Heidegger, 1995, 310, 337-338

276. Heidegger, 1995, 339

277. "Philosophy is philosophizing" (Heidegger, 1995, 4)

278. Midgley, 1996, 13

279. Midgley, 1996, 13

280. Mill, 1859/2010

CONCLUSION AND THE ROAD AHEAD

1. Foucault, 2006, 151

2. Bookchin hinted at this role of freedom and argued that it renders invalid all ideologies that suggest formulas to "predetermine our behavior irrespective of our wishes" (Bookchin, 2005, 241).

3. This is, perhaps, how we should understand the seemingly paradoxical quote from Bashevis Singer: 'We must believe in free will. We have no choice" (Goodman, 1987).

4. See, for example, Yudanin, 2013

5. J. S. Taylor, 2005, 18-19

6. See, for example, Beauchamp, 2005, 314-316, 323, 325-326. Similar reasoning can be found in Kant as well.

7. Timofeeva, 2018, xi

8. The natural right theorists deem certain rights inalienable, belonging to their bearers by virtue of the latter's capacities. This way, an agent who is capable of expressing her thoughts would have the right to free speech, no matter what are the conventions of the society in which she lives. The first germs of this concept can be seen in Heraclitus who declared that "all human laws are nourished by the one divine law" (Diels, 1903, 82, DK22B114), yet here the link between the ability of the agent and the corresponding right is not yet pronounced. Cicero's *vera lex*, which is "right reason in agreement with nature," is more pronounced as congruent with the nature of the agent as "we need not look outside ourselves for an expounder or interpreter

of it." (Cicero, 1928, 211 (3.22)). It is Locke who first explicitly ties the rights to the abilities of their bearers, as "all that share in the same common nature, facilities and powers, are in nature equal, and ought to partake in the same common rights and privileges" (Locke, 1689/1988, Book I, Chapter VI, §67 (location 936)) – a conception later enshrined in political documents like the US Declaration of Independence, the UN Universal Declaration of Human Rights, and more. Rousseau suggests that, since non-verbal animals partake in a measure of our nature, they ought to partake in a measure of right – the right that is based on nature, or natural right (Rousseau, 2002, 84-85).

9. Jonas, 2001, 108-127

10. For an introduction to neural networks, see Graupe, 2013, esp. Chapter 1, 1-3, and Chapter 3, 1-16. See also footnote 71.

Index

About the Author

Michael Yudanin's primary interest is in the philosophy of freedom and its evolution. He has published papers on freedom and liberty, as well as on Kantian ethics and its limitations.

www.ingramcontent.com/pod-product-compliance
Ingram Content Group UK Ltd.
Pitfield, Milton Keynes, MK11 3LW, UK
UKHW012118150926
475560UK00002B/175

9 781793 620187